# PRAISE FOR *Farming for the Long Haul*

"Michael Foley's passion and his lifetime of research and lived experience comes through in this primer of agricultural history and personal philosophy that is sure to prompt an important dialogue about the future of agriculture and the political economy. Agriculture is a shared expression of who we are, and I believe we need more people thinking deeply, questioning, and sharing their insights. This book provides many tools and references to ask informed questions and encourage a richer discussion about progress."

—DORN COX, farmer;
founding member of Farm Hack

"Globally, humanity urgently needs to transform the way we make our livelihoods if we're to thrive into the future. In *Farming for the Long Haul*, Michael Foley shows that food and farming are at the heart of this, and he gives us some fine tools for rethinking them. There's a heft to his book that speaks of hard work—both in the study and on the land—but there's also a lightness to the writing that makes it a pleasure to read. The world badly needs more farmer-scholars like Foley."

—CHRIS SMAJE,
Small Farm Future, Somerset, UK

"This book is a modern peasant's manifesto! Small farmers today have a stark choice, Michael Foley tells us. We can either buy into the current agricultural and food system and fail utterly, or we can try to change the system entirely. This book outlines the history of agriculture, shows where we've gone wrong, and recounts the practices and values of the most resilient long-haul farmers throughout the world. Then Foley sets up a visionary solution aimed at helping small farmers both survive the dwindling stages of our current system and position themselves for the dramatic changes the future holds. *Farming for the Long Haul* extends, expands, and updates Wendell Berry's *The Unsettling of America*, then puts forth a vision of a land of resilient small farms ready to survive the present and thrive into the future."

—CAROL DEPPE,
author of *The Resilient Gardener*

# Farming for the Long Haul

## *Resilience and the Lost Art of Agricultural Inventiveness*

MICHAEL FOLEY

Chelsea Green Publishing
White River Junction, Vermont
London, UK

Project Manager: Sarah Kovach
Project Editor: Ben Watson
Acquisitions Editor: Michael Metivier
Copy Editor: Laura Jorstad
Proofreader: Caitlin O'Brien
Indexer: Shana Milkie
Designer: Abrah Griggs

Printed in Canada.
First printing January, 2019.
10 9 8 7 6 5 4 3 2 1 19 20 21 22 23

Our Commitment to Green Publishing
Chelsea Green sees publishing as a tool for cultural change and ecological stewardship. We strive to align our book manufacturing practices with our editorial mission and to reduce the impact of our business enterprise in the environment. We print our books and catalogs on chlorine-free recycled paper, using vegetable-based inks whenever possible. This book may cost slightly more because it was printed on paper that contains recycled fiber, and we hope you'll agree that it's worth it. *Farming for the Long Haul* was printed on paper supplied by Marquis that contains 100% postconsumer recycled fiber.

Library of Congress Cataloging-in-Publication Data
Names: Foley, Michael W., 1945- author.
Title: Farming for the long haul : resilience and the lost art of agricultural inventiveness / Michael Foley.
Description: White River Junction, Vermont : Chelsea Green Publishing, [2019] | Includes bibliographical references and index.
Identifiers: LCCN 2018037034| ISBN 9781603588003 (paperback) | ISBN 9781603588010 (ebook)
Subjects: LCSH: Farms, Small--United States. | Farmers--United States. | Agriculture--United States.
Classification: LCC HD1476.U6 F65 2019 | DDC 338.10973--dc23
LC record available at https://lccn.loc.gov/2018037034

Chelsea Green Publishing
85 North Main Street, Suite 120
White River Junction, VT 05001
(802) 295-6300
www.chelseagreen.com

# Contents

# Chapter 1

# Farming in the Ruins of the Twentieth Century

This is a book about farming today with an eye to our uncertain future. It is about building a viable small farm economy that can withstand the economic, political, and climatic shock waves that the twenty-first century portends. Uncertainty and periodic disruptions have always been part of the background as humans worked to bring the fruits of nature to the table. In the twentieth century the economics of a growing urban, industrial society wrecked the dream of a smallholder society on which Thomas Jefferson bet his hopes for a democratic America. Economic hardship and the years of drought that ended in the Dust Bowl wrecked the dreams of millions who thought to live off the land, sustaining their own livelihoods while supplying the burgeoning cities of an industrializing giant. And they continue to take their toll on hard-nosed agribusiness people and idealistic young farmers alike as we make our shaky way into the twenty-first century. The shocks of the coming decades arguably will be greater—unprecedented in the short history of industrial agriculture but familiar enough in the history of descent of other civilizations.[1]

The next chapter recalls some of our recent history, but we will not dwell on a critique of the industrial model of agriculture as it developed over the last hundred years in this country. That has been done ably by many before me.[2] The road to the present impasse is part of a recent past that should inform our decisions about how we farm and what we can expect from farming today. In fact, too much of the farming advice that new farmers are receiving today ignores that sorry history and repeats its mistaken assumptions. It ignores as well the real perils farming faces in the wake of the ruin that twentieth-century agriculture, and the fossil fuel economy that supported it, has wrought. And too often missing in our thinking about the future of farming is an awareness of the long history of resilience and adaptation that other farming cultures brought to bear as they faced other disasters, persisted, and found new and better ways to secure a livelihood and live sustainably upon the land.

This book looks cautiously forward toward the challenging prospects of the next several decades, trying to suggest ways to prepare a farming future that can survive the transition to a future very different from the present. And it will do so drawing on a still-deeper past, recalling examples from the vast history of agriculture and the endless array of agricultural adaptations that farmers have devised, gleaning lessons from more resilient agricultures, ones that withstood not just the vagaries of the modern capitalist market, but environmental and social shocks of the past that rival what lies ahead.

Successful farming societies—and there have been many of them throughout the history of agriculture—have never been driven by the profit motive, though profit has played a part in some societies, at least where farmers were free and markets significant. Instead, they have been governed by principles of resilience—by the effort, to put it simply, to survive and to do so well under uncertain conditions. Those principles include: a dedication to their own subsistence, above all; sophisticated approaches to management of the natural resources on which farming depends; and dependence upon a vibrant local community as a first line of defense against calamity.

Contemporary efforts to teach the latest generation of farmers the business skills necessary to turning a profit fly in the face of this history, just as they ignore the short, sorry history of profit-oriented farming best on display in the United States. Often these efforts are based on a misunderstanding, a subtle morphing of two disparate notions. "You can't farm sustainably," many say, "unless you're economically sustainable." True enough. But then comes the subtle wrong turn: economic sustainability = profitability. Economic sustainability in the historical record, however, relies more on resilience than on profitability. In fact, the quest for profitability can be the enemy of resilience, as when American farmers cut the woodlot, eliminated the milk cow, the hogs, and the chickens, and took down the barn—the better to farm fencerow-to-fencerow in the quest for profits in markets fixed against them. Resilient farmers know the difference between short-term profitability and long-term economic sustainability. Farmers condemned by debt and bad advice to constantly track the bottom line generally do not.

## Millennia of Examples

Farming societies—tribal, peasant, yeoman farmer—have often been prosperous. Why, then, the prejudice that farming is a losing game, peasants forever condemned to misery and ignorance, tribal societies a dead end? No doubt the reasons are many. History remembers disaster better than success, and military prowess over the everyday persistence of ordinary people. Peasants don't easily raise armies and thus are easily overrun by marauders on horseback, who combine military advantage with terror, slaughtering men, women, and children, burning crops in the field, driving off the livestock. And when peasants do raise armies, they are often defeated, if not by their enemies then by the likes of the communist cadres who assumed leadership in twentieth-century peasant revolts around the world and who reshaped peasant life and society in the wake of victory.

## But Weren't the Lives of Traditional Farmers Nasty, Brutish, and Short?

In a word, no. A misleading version of history tells us that before the twentieth century, the average life expectancy was somewhere between thirty-five and forty-five years, depending upon where and when you lived. But those are *averages*. They include the large proportion of the population who died before their tenth, or even fifth, year. Both historical records and studies of hunter-gatherer cultures show an infant and child mortality rate of about 50 percent around the world. So if the average life expectancy of a society is forty years, then the life expectancy for an adult who survived childhood would be in the seventies. In fact the life span of adult hunter-gatherers studied in the twentieth century was sixty-eight to seventy-eight years, similar to the post-childhood life span of an adult in eighteenth-century Sweden.[3]

Infant and child mortality was reduced drastically starting in the late nineteenth century with public health measures around clean water and sanitation, and over the twentieth century these efforts, combined with inoculations against common infectious diseases, spread around the world, increasing life expectancy everywhere. But life expectancy in traditional cultures was relatively robust for those who lived to age ten or twelve, and a few traditional societies saw large numbers of people living to a hundred years or more.

We also tend to remember the recent conquest and absorption of so-called primitive societies by the West rather than their persistence over hundreds and thousands of years. Tribal societies for the most part could not compete with colonial newcomers, neither militarily nor culturally. Traditionalists like Tecumseh or Crazy Horse found themselves fighting losing cultural battles against the majority of their own people enthralled by the stuff the industrial-based intruders had to offer. The success of these societies in getting a subsistence and building a rich culture over millennia came to look, from this perspective, like backwardness.

Since the dawn of the republic, moreover, American farmers have been forced to sell on a market manipulated by governments and speculators, railroads and multinational corporations. They are "price-takers," as the economists put it, not price-makers. We remember the revolt of farmers over the terms of that market in the post–Civil War era but not the prosperity of the rural economy that followed, the age that built the lovely towns that dotted our countryside in its aftermath. We see only the ruins of those towns in the wake of the seventy-year advance of the industrial model. We remember the struggles of homesteaders unsuited to the task of learning to farm a new terrain or condemned to failure by the attempt to bring traditional farming methods to the arid West. And we know that farm country has grown increasingly deserted over the last seventy years. But we don't recognize the achievements of the settled agriculture of the nineteenth and early twentieth centuries in many parts of this country, and we ignore the undermining of those achievements through government policy, expert advice, and growing monopoly power in the marketplace.[4] Too often, and too persistently, we blame the victim.

We have much to learn from the social organization, farming skills, ecological wisdom, and political resistance of the farming societies of the past. In place of peasants and farmers, economists and policy makers have elevated the agribusinessman. But agribusiness as we know it is unsustainable even for many agribusinessmen, and certainly for the soil and our food system. The few who succeed in *that* losing game are scarcely models for the agriculture of the future. In seventy years of ruinous farming practice, industrial farming has exhausted our soils, poisoned groundwater and a large portion of the Gulf of Mexico, and provided the basis for a food culture that is making most of our population sick. To go forward we will have to look back and attempt to recover, with the wisdom of hindsight, the lessons that traditional agricultural societies have to teach about subsistence, stewardship, social organization, and economic and political resilience.

Food systems are not just about the production, distribution, and processing of foods. They represent distributions of land and power, wealth and poverty, plenty and famine. As farmers we can buy into the food system we are born to and try to rise to the top in a usually losing game, or we can try to change the system and the practice of farming itself. The contemporary sustainable agriculture movement is an attempt to change one very complex and powerful food system. But we will fight against odds of our own making, and not just those devised by that system, if we accept its premises, if we suppose there are no alternatives to struggling to make it in the market economy. And there is a third option: We can return to the norm of the successful agricultural societies of past centuries that were based on a strategy of subsistence first, resilience in the face of both natural and human-made challenges, and continuing invention, and grounded in the reciprocity and fellow feeling of living communities. Understanding the alternatives that human societies have devised in the past is essential to enlarging our vision of farming for the long haul.

## Who Are We?

Farming is a vocation, Wendell Berry insists, not a job. Starting there is already a step toward resilience, because it means we recognize that we are in this for the long haul and that it will take a great deal to drive us out. I knew farmers in southern Maryland who clung so desperately to farming that they carried on even when it made no money and little sense, renting land so they could run tractors on it and plant, spray, and harvest corn and soy on weekends while the rest of their lives were taken up with the nine-to-five jobs that paid the bills. Only the Commodity Credit Corporation kept their hobby from bankruptcy. To one degree or another most farmers in the country are like those farmers. And even those who farm more sustainably and with less reliance on government handouts do so out of their dedication to farming as a vocation.

Farmers, moreover, may or may not have much of the entrepreneur in them, but even if they do it is not the mentality of the entrepreneur whose happiness is building a company, no matter what its product, and squeezing a profit out of it. No, a farmer builds a farm for the sheer pleasure of working the land and working with animals. The *business* of farming may or may not be of interest to farmers, as much as it has to be attended to in any case. Farming is not first of all a business, as our business advisers like to insist. It is a way of living. Either it satisfies a passion in us, a deep need, or we will pretty quickly give it up.

Those facts guarantee a sort of personal resilience, but they don't guarantee that anyone with the requisite passion will be able to continue to farm. Nor do they guarantee that the style of farming adopted will be one that has a future. Those outcomes depend on skills and attitudes and strategies that will be the topic of further chapters.

We can learn a lot about ourselves and the tasks we face by recognizing that we farm because we love to. What that means concretely will vary from farmer to farmer. But any open discussion among farmers reveals some near constants: a passion for nurturing animal and plant life; a love of working outdoors; an appreciation for the mysteries of nature; a commitment to feeding people; a talent for troubleshooting; a fierce independence. Each of these tells us something about how farmers go about their work as opposed to, say, engineers or manufacturers. We nourish, we don't make. Even as plant or animal breeders, we bow to the mysteries and the intransigence of nature. We shape our environment, but if we are wise we don't overturn it. We depend upon nature and natural processes and must work with them. We are multifaceted doers, as capable of rigging an irrigation system as managing a thousand transplants or moving a herd of goats. And we want to do it ourselves, even as we recognize the need to work cooperatively with others. These generalities may apply more to small, alternative farmers than to midwestern corn and soy farmers, but to one degree or another they likely apply to all.

These characteristics in themselves add up to a lot of farm resilience. And that resilience explains how many farmers and their farms survive year after year, even under the present, very unfavorable circumstances. It takes very bad advice, and very bad investments, to put a farm under. That fact speaks to the enormity of the bad advice that has been doled out to farmers over the last century.

The pages that follow attempt to correct that bad advice, not with new advice about how to farm, what to grow, or how to market it—there is plenty of good advice out there on such topics—but with strategies for farming in both hard times and good, in market economies and outside them, and reflections on the social context for farming well and over the long haul.

## What Will the Long Haul Look Like?

The chances that the next fifty years will not bring enormous, not to say cataclysmic, disruptions to our way of life are nil. World oil reserves will be exhausted in as little as fifty years.[5] Ditto natural gas. As will the lithium that powers today's most sophisticated batteries. As this suggests, transportation is equally imperiled. If the world doubled the number of electric vehicles on the roads every year, it would take until 2035 to surpass the total number of gas guzzlers on US highways alone; currently electric vehicles—and projected storage for intermittent renewable energy generally—depend on lithium batteries. There is no current production line for electric pickups, tractor-trailers, heavy equipment, or farm tractors. And for those who still imagine that the electric tractor will replace gas- and diesel-fueled versions, another, more dire limitation arises: At current rates of erosion, the world's topsoil will be gone in sixty years.[6] In short, the large-scale agricultural and food delivery system as we know it has at most a few decades before it exhausts itself and the planet with it.[7] Only radical innovation can make possible production on a scale that could hope to feed the world's population.

Climate change poses still more disturbing possibilities. We could halt all greenhouse gas emissions today and climate change would continue to accelerate. Sea levels are already rising; the Arctic ice cap is evanescent in the summer and thinning each winter. The Antarctic glaciers are calving at an alarming rate. And exposed permafrost is already emitting methane gas, a far more potent greenhouse gas than $CO_2$, increasing warming trends no matter what the signatories of the Paris Agreement do or don't do. Coastal flooding is already producing climate refugees, as are the drought and drought-exacerbated wars of Africa and the Middle East. And devastating storms like Katrina in 2005, and Harvey and Maria in 2017, are becoming more frequent.

Disruption will affect different regions and corners of the world differently, but we will all be affected. In the United States the arid West and the wet Northwest are both expected to become significantly more hot and dry. The Midwest and Northeast are already experiencing warmer, wetter weather, with severe storm events increasing. Catastrophic flooding like that experienced in Houston, Bangladesh, and India in 2017 is likely to increase in coastal zones and elsewhere, and heat waves across the world are becoming hotter. Food production is likely to be affected more profoundly than other economic activities. Farmers will be called upon to produce under increasingly adverse environmental conditions with less and less access to the sort of inputs that made the twentieth century's surge in food production and population possible. And they will have to produce with more and more limited access to markets, with greater transportation costs, and in the face of rapidly rising demand as conventional world production slumps or is cut off. Markets for local food may flourish, so long as local communities are flourishing. But the end of the oil economy, combined with the economic impacts of climate change, will result in fewer and fewer opportunities to flourish beyond the local market for any but the most ruthless speculators.

None of this should be controversial, and most of it has been predicted for decades. As the Club of Rome pointed out in 1972, we live in a world of limits, and sooner or later those limits will rise up to bite any civilization that insists otherwise. Of all people, farmers should recognize the dilemma posed by the limits that nature sets us, however often the agricultural experts and agribusiness suppliers insist there is no end in sight. As the Greenhorns have put it, "We face a dystopian future, with guaranteed-unpredictable weather, the impending collapse of the fossil fuel economy, endlessly consolidating monopolies, and a country that is, for the first time in our history, majority urban."[8] Over the next decades we will be farming "at the end of the world," that is, the world as we have known it, the world that virtually every institution in our society continues to regard as the only guide to our future. That it isn't and shouldn't be our guide is the premise of this book.

## How Then Should We Live?

We may be convinced that business as usual will not serve us for long, but all we have to work with is uncertainty. In the face of limits, we should probably try to stop pushing those limits, but there is much we cannot do to anticipate an unknown future. What we *can* do is build resilience into our lives and adaptability into our farming. Permaculturalists who have faced up to the limits to growth justify their recourse to backhoes and bulldozers by saying we need to use what we have while we have it. They're probably right, if what we use such tools *for* is building the sort of resilient farms that have a chance in that uncertain future.[9] And we can argue that burning fossil fuels for farming can be justified if we're also finding ways to sequester that carbon. But these are questions about the means and morals of building resilience. How we should live in the face of dire prospects is a bigger question.

We don't know what future farming will look like, but if we have a passion for farming, for caring for a piece of land and even reclaiming it from the destruction past human uses have wrought upon it, then we start by doing what we love—what we love and what we know is right. As Wendell Berry often insists, we can't save the future; we can only do what is right for the present. And as Berry also insists, what is right must be right for the land and the people of the land.

If we have a passion for farming—for producing food not just for ourselves and our families but for others on a sufficient scale to feed them for a time—then we have to farm for production's sake. Almost inevitably, in today's circumstances, that means farming for a market of some sort. We may want to build all sorts of reciprocity into that—barter, exchange, giving. But above all, for the long haul and for right now, we need to build resilience, and that means we have to build a farm that makes economic sense in the here and now.

Doing so, however, does not necessarily throw us back into the sterile logic of the marketplace, the one all those farm advisers venerate. "Economic sense" will depend upon the circumstances that each of us brings to farming: other sources of income, other vocations, our hold on the land we work, and so on. And "economic sense" should include our own livelihoods, not just the profits, and perhaps wages, we wrest from the market. Here is how Berry accounts the success of his "marginal" farm: "As income I am counting the value of shelter, subsistence, heating fuel, and money earned by the sale of livestock." Thus, "once we have completed its restoration, our farm will provide us a home, produce our subsistence, keep us warm in winter, and earn a modest cash income. The significance of this becomes apparent when one considers that most of this land is 'unfarmable' by the standards of conventional agriculture, and that most of it was producing nothing at the time we bought it."[10] Such a farm may or may not produce all the income one currently needs, but it is a farm that produces resilience. It is a farm that will be better equipped to survive the uncertainties of the market economy, may creatively address the implications of climate change, and could provide one small contribution to the resilience of the larger community.

## What Is Resilience?

In the jargon of systems theory, *resilience* is often defined as "the ability of a system to absorb disturbance and re-organize while undergoing change so as to still retain essentially the same function, structure, identity, and feedbacks."[11] A "system" can be an ecological niche, a watershed, a forest, a farm, a business, a community, a state. To the degree that an identifiable physical or organizational space is strung together out of interrelated parts, it is a system. That stringing together, however, may be brittle or ductile, fragile or durable. Change may tear it apart or be absorbed by it. It may fragment into subsystems or disintegrate and degrade. *Resilience* describes the qualities that will help it persist, and even thrive, in the face of drastic change.

So we may attribute resilience to an ecological system, or we can find the system degraded, even broken, because of its inherent fragility or because of the violence of some intervention, human or not. We attribute resilience to a piece of land that can absorb water, fire, drought—as most deserts and some forests do—and spring back to new life on much the same terms as the old. Or we may bear witness to its degradation and attest to the poverty of its present state in comparison with what came before.

The same is true of a farm or a food system, with the added complication that this is a system also meant to serve human uses, subject to human management. A farm cannot be resilient unless the land itself is resilient. But the farming enterprise, too, must be capable of dealing with drastic disturbances; it must be adaptive. And that means that the people who farm must themselves be resilient—that is, they must be able to reorganize in the face of negative feedback or even collapse; they should be able to bring new and old ideas and techniques to new circumstances; and they must do so on a trial-and-error basis, making choices that can contribute to continued resilience. The more we know of our past, the better prepared will we be to make wise choices in light of the challenges we face.

Like every farming culture before us—indeed every culture before us—we are in this for the long haul. Again, we have no better guide than Wendell Berry. Recounting one of his own drastic mistakes on his land, Berry writes,

> The true remedy for mistakes is to keep from making them. It is not in the piecemeal technological solutions that our society now offers, but in a change of cultural (and economic) values that will encourage in the whole population the necessary respect, restraint, and care. Even more important, it is in the possibility of settled families and local communities, in which the knowledge of proper means and methods, proper moderations and restraints, can be handed down, and so accumulate in place and stay alive; the experience of one generation is not adequate to inform and control its actions.[12]

Farming systems throughout human history and prehistory have achieved the resilience that follows from such a culture, and many of them maintained it for hundreds, even thousands of years. Ours has not. Our task today is to move farming back toward the norm, toward a greater resilience in a time of imminent stress. Our task is to reestablish the bases for farming for the long haul ahead.

## Chapter 2

# A Short, Unhappy History of Business Advice for Farmers

In the 1920s the newly minted profession of agricultural economist plied its wares among America's struggling family farmers. The farmers had profited from cheap land, thanks to the 1862 Homestead Act, and high commodity prices, thanks to the First World War. In 1918 commodity prices collapsed. In the upper Plains states, recently settled by waves of homesteaders, including my great-grandmother, the problem was compounded by the beginnings of drought conditions that would eventually become the Dust Bowl of the 1930s.

Savvy economists and their friends in banking concluded that the failure of so many family farms lay in the farmers' lack of business sense. Extension agents, trained by university-bred agricultural economists, traveled the country teaching farmers how to keep books, analyze a balance sheet, and choose wisely among competing crop options. They also pushed mechanization, encouraging farmers to expand and adopt the newly developed tractors in place of horse-drawn agricultural implements. Not many farmers bought. They knew horses. They didn't know the new machines, and they were all too aware of how unreliable they were. Besides, buying a tractor or combine required borrowing. The bankers and economists were eager to counter the resistance to indebtedness, but most farmers were not interested.

The tractor companies mostly moved offshore, selling more tractors in the 1920s in the Soviet Union, where the government was eager to apply the latest American technology, than at home. Some of the extension agents went to work for the Soviets, too, helping them devise the enormous state farms that they, along with their communist sponsors, thought the only solution to farming on a modern scale. A few such experiments were tried in the United States, without the coercive powers of the state to back them, but they almost uniformly failed.[1]

In the Plains states, homesteaders—encouraged by agricultural experts and politicians to strike out west with the promise that "rain follows the plow"—went under in the growing drought conditions of the 1920s. Many of the "progressive" farmers who bought into the new economic logic failed, too, though California agriculture, already big enough, expanded production and markets in the run-up to the Great Depression. But it wasn't until the Second World War brought higher commodity prices that the farmers who survived once again flourished.

Most had long since fled the countryside, like my great-grandmother Sarah McCann, who homesteaded outside of Lewiston, Montana. As the farm economy failed, she moved into town, which would remain a boomtown for a few more years, and then to another boomtown, Roundup, Montana, where my father was born. But no one remains from that clan. All moved on. Lewiston has recovered a little of its former wealth. Roundup still struggles.

That first generation of eager, economically trained advisers set the pattern that would persist through several cycles of boom and bust for American farmers. And as their advice took its toll, the countryside has emptied, gradually at first, then more rapidly as the industrial model of agriculture laid hold of the land. Following the Second World War, American farmers finally bought the promise of the new farm machinery and the advice of the economists. Many young men from farming families had learned how to handle heavy equipment in the battlefields of Europe. Those who stayed home had tinkered with their autos long enough to feel comfortable with the internal combustion engine. And farm machinery was cheap. Postwar reconstruction devised in Washington by the country's leading industrialists saw to that. So was credit, another feature of postwar economic planning. Farmers expanded, at first slowly, and then, as farm prices continued to fluctuate, driving indebted farmers from ancestral homes, the survivors grew more rapidly, buying up farmland in an attempt to "scale up" to beat falling profit margins and high risk. By the 1960s policy makers were saying there were "too many farmers," and American farm policy inexorably whittled down those numbers, taking rural communities along with the abandoned farms.[2]

The logic of the agricultural economists, already evident in the '20s, was spelled out in the early '70s by then secretary of agriculture Earl Butz, whose contribution to the lexicon was his admonition to "Get big or get out." Food would become a weapon in the Cold War, and American farmers would be the cannon fodder in the quest for dominance in international grain markets. In Butz's view and that of the farm advisers, getting big meant doing your cost-accounting, identifying the profitable crop or two, preferably wheat, corn, or soy, and putting everything into it.

In the course of the 1970s (when the deficiencies of Soviet agriculture created a boom market for American grain), farmers took out fences and hedgerows, cut the woodlot, let the vegetable garden go to grass, drove out the chickens and pigs and family cow, and tore down all the outbuildings that went with them. All those "uneconomical" ventures had to make way for rows upon rows upon rows of corn or soybeans, wheat or oats. The aim was now to farm fencerow-to-fencerow. Depopulated of farmers, the land was now depopulated of farm animals. Farm wives learned to buy their food at the new supermarkets and look for supplemental income, increasingly necessary to sustain a middle-class farm lifestyle, hopefully with a good government job. No more churning butter and collecting eggs for sale in town. No more growing and canning food for the winter in the family vegetable garden. No more woodstoves fueled with farmstead wood. Fossil fuels were more "economical."

The new economies didn't avail much. The next downturn in commodity prices came in 1979, the same year that certified economic genius, Federal Reserve chair Paul Volcker, jacked up American interest rates, wrecking the world economy. Farms failed in the tens of thousands, saddled with debt by trying to get big to avoid getting out. Between 1979 and 1985 fully half of American farms went under. American farmers had the options of watching the sheriff foreclose on the family farm, shooting their bankers, or shooting themselves. Most chose the first option. A few chose one or both of the latter. In any case, farmers were the losers. But they were doing what the experts said. They had gotten bigger, and now they were getting out.

## Spreading the Way of American Farming

The story was, if possible, more tragic in what we used to call the Third World. The Green Revolution of the 1970s propagated by American crop breeders, the Rockefeller Foundation, and the World Bank encouraged farmers in Mexico, India, Thailand, the Philippines, and elsewhere to adopt new, highly productive crops developed by American researchers and their colleagues around the world. The new crops only needed abundant and reliable sources of water, fertilizer, and pesticides. And their introduction indeed rewarded the hopes of their promoters. Production of basic commodities, principally rice, wheat, and corn, soared in adopting countries. The progressive farmers who seized upon the promised opportunities were lauded by agricultural extension agents and their governments and celebrated in the press.

But the boom didn't last. Rapidly increased production drove down commodity prices, making the expense of the new techniques increasingly hard to manage for the average farmer (and many poorer farmers never participated). In the Punjab and Rajasthan regions in northern India, aquifers were pumped dry, traditional waterworks left to decay, and much of the land desertified by a combination of soil-depleting fertilizers and drought. In recent years thousands of indebted farmers have committed suicide drinking the pesticides they no longer need on their parched land.

In Mexico peasant farmers were coerced into adopting the new crops by monopoly buyers and rewarded with cheap fertilizer in exchange for political loyalty to the regime that promoted the new technology. Already by the 1980s peasants reported that *la tierra ya no da* ("the land is exhausted") and they rightly pointed to the artificial fertilizers to which they had become addicted. The worst blow came in the mid-1990s when NAFTA (the North American Free Trade Agreement), which economists promised would bring prosperity to Mexico, ushered in a flood of cheap US corn, undermining the little cash flow that Mexican peasants had been able to count upon. Millions crossed the border, where cheap labor continues to suppress the prices of the fruits and vegetables upon which a new generation of farmers is hoping to depend.

## What the Economists Don't Know

The economists should have known better than to throw farmers on the mercies of the market. Farmers, they all acknowledge, are price-takers not price-makers. But the economists believe in a self-equilibrating model of market behavior. Alas, the economists' God (that is, the Model) gives no thought to the wrecked lives, lost topsoil, or abandoned communities that the market leaves behind in the process of "reaching a new equilibrium." All those lost farmers are just so much fodder for the efficient market game.

Is there an alternative to the economists' straitjacket advice? Certainly it's a good thing to keep the books, minimize expenses, and maximize sales. It can't hurt, either, to be aware of the profit margins available for different crops and enterprises on the farm. But are we condemned to the sterile—and too often disastrous—logic of cost-accounting?

The Russian agrarian sociologist and economist Alexander Chayanov found that peasant economies were based on a more complex economic calculus. Though peasant farmers around the world produce for the market, they also produce for themselves. Their production expands and their income rises not as their business sense increases, but as their families grow. And once grown and provided for, peasant farmers scale back. Throughout the world, subsequent studies have found that the attitude of traditional farmers is "subsistence first": Provide for household food needs, turning to the market as need and opportunity allow. Chayanov was executed by Stalin in the 1930s for insisting that, given that logic, the large-scale collective and state farms Stalin favored would fail. (And fail they did.) Communist and capitalist farm advisers in the 1920s shared a disdain for the peasant household economy, and the attitudes of Marxist and neoliberal economists haven't changed much since then. The traditional subsistence-first attitude has driven agricultural advisers to distraction, from British and French colonial agents to the US Agency for International Development (AID) advisers and even Peace Corps volunteers, as traditional farmers reject the advisers' agricultural innovations, including supposedly more lucrative crops, if adoption threatens proven strategies of providing for their families. The next chapter will explore the lessons that the subsistence-first agricultures of peasants, indigenous peoples, and traditional American farmers have to teach us about genuine economic sustainability.

We all have to pay the bills, no doubt. But the woodlot, vegetables and fruit from our market gardens, that flock of chickens or ducks, those pigs and milk animals are also part of the value calculation, as Chayanov showed. It makes sense to cling to them in a market-driven world, because they provide the security and resilience that the market cannot and will not. It might also make sense, of course, to compare the costs of maintaining animals with the costs of the same goods bought at market; but we all know that our chickens' eggs are much more healthful than the supermarket variety, "free-range," "organic," or not. These aren't trivial considerations, as the present state of American farming and American health demonstrate. And they can make the difference in hard times between survival with dignity and destitution. Wendell Berry quotes an account of Kentucky in the 1930s that underlines how little the state was affected by the Great Depression. In fact, people returned to the land and their homes, as Berry explains, "because at home they still had families who were growing a garden, keeping a milk cow, raising chickens, fattening hogs, and gathering their cooking and heating fuel from the woods. Now," he adds, "eighty years and much 'progress' later, where will the jobless go? Not home, for there are no 1930s homes to go to."[3]

There is something else missing from the cost-accounting of the business experts. A character in one of Wendell Berry's novels recalls working on every farm down his road, and he's proud to say that not a penny was exchanged. Cooperative labor arrangements were common in American farming almost to mid-twentieth century, before large-scale fruit and vegetable production on the California model became common. Harvesttime brought out everyone—farmers and their workers, farm wives, town wives and their children, even college kids home from school for the weekend. Orchards were harvested, tobacco sorted, tomatoes brought in. Schoolchildren would take a couple of weeks off in the Santa Rosa, California, area just for the prune plum harvest. A scheduled school break for the potato harvest continues to be part of the school year in Aroostook County, Maine. And Amish communities have always come together around harvest chores.

The newest farmers today, and most of the small farm owners of all ages, continue to rely on volunteer labor. Collective work parties are part of the new ethic of solidarity among young farmers. Sharing resources—traditional in the older American agriculture—is embodied in the structure of the Greenhorns and California's new Farmers Guild. And sharing knowledge is part of the farmer and farm intern exchanges of the Collaborative Regional Alliance for Farmer Training (CRAFT) movement.[4] Our business advisers would prefer we become employers and leave the schooling to accredited colleges and universities, the training to extension agents and other certified experts. But the newest farmers know the difference between the classroom and the field and have often sought out field experience on a variety of farms before starting up their own operation. And they are eager to continue to educate the aspiring farmers who come behind them and are happy to share their own labor with their neighbors.

The values that small farmers often embody are outside the logic of cost-accounting. They recall a time and a society that was more resilient than ours, that did a better job in most cases of taking care of its own, and that provided for most everyone in hard times and good. Not a perfect society, to be sure, but one where frugality, self-sufficiency, and community embodied the wisdom of the Chayanovian peasant, teaching that we will only survive the vagaries of the market if we provide prudently for ourselves and share with others.

The best of the new business advisers at least reject the logic of monocropping and mindless scaling up. They can offer good advice for the market-oriented.[5] But we can cost out everything and still come up losers. The winners will always be the bankers, tractor manufacturers, chemical and seed companies, and, of course, that legion of agricultural advisers, including business advisers, so willing to help us out. When farmers fail these folks go on with their jobs, which have always paid more than the average farmer earned in any case. For farmers "acquiring business skills" has too often left us with debt, diabetes, and the death of our dreams.

The next generation of farmers can do better than that, provided they take all that advice with a large grain of salt and seek out examples of genuinely sustainable agriculture to guide the building of a new farming culture. Sustainability doesn't consist exclusively of generous doses of compost and a few hedgerows. And economic sustainability doesn't simply equal profitability in our rigged markets. Those ingredients are important, but they are not enough. We need, among other things, an economics that takes account of how we live from the land even when the money economy fails us, as it surely will. We need resources to sustain us in good times and bad, not only on the farm but in our community. And we need tools to confront the challenges that climate change and the end of easy energy will bring—are bringing right now.

The chapters that follow offer an alternative model of economics for the farm and another sort of advice than that of the farm advisers. They offer advice or, better, examples based in the millennial experiences of indigenous farmers and peasants and carried on even in the United States before the industrial transformation of agriculture in the last century. These are agricultural societies that persisted in some cases over centuries and that, when confronted with disaster and environmental degradation, adapted, inventing new ways to live more sustainably with the resources that remained. We can learn a lot from the latest research and find suitable uses for the new tools that our society continues obsessively to churn out. But we can learn much more that will sustain us and our livelihood on the land over the next decades from the examples of past and present agricultures founded on appropriate technologies, long-tested practices, economic resilience, and cultures of restraint and stewardship.

# Chapter 3

# Subsistence First!

Peasants, indigenous agriculturalists, and old-time American farming families farmed first and foremost to feed their own families and those in need in their communities—only secondarily, if at all, for a market. They may have practiced shifting agriculture or were settled permanently in villages; they may have been members of free, "primitive" or "tribal" societies, or peasants tied to a manor. They may have produced for a lord or the king or a market. They may have been "homesteaders" or market farmers. But all produced first of all to sustain their own families and dependents. And by and large the social order in which they lived supported that intention, coming to the aid of the weaker members in times of stress, extending credit to the desperate farm family, or refusing to exact more in tribute or manorial dues than familial and communal subsistence allowed.

As mandatory participants in the contemporary market economy, today's farmers have to produce for the market as part of their own subsistence strategy. Even for us, subsistence comes first. If we can't provide for ourselves, we can't feed others. Somehow or other we have to stay afloat economically. We tend, however, to see the issue in purely monetary terms. If we don't earn enough money—on the farm or in off-farm employment—we don't stay afloat. We share with all farmers in the past our determination to subsist, one way or another. But traditional subsistence practices have relevance beyond that abstract commonality, because traditional practices were built on strategies of resilience. A resilient farm is one where the farm family can draw on its own production to feed itself in hard times, when markets fail or market prices prove ruinous. That's one reason why the business advice of the past has been so disastrous. With nothing to fall back on, completely dependent on the grocery store and utility company for the necessities of life, American farmers bilked by the market, short on cash and over their heads in debt, fled the countryside by the millions over the last one hundred years.

We can't escape the market today. Nonparticipation is not a reasonable choice. As John Michael Greer points out, "No one is actually forced to participate in the market economy in the modern industrial world. Those who want to abstain are perfectly free to go looking for some other way to keep themselves fed, clothed, housed, and supplied with the other necessities of life"; but they are apt to find their way blocked by law and public opinion, not to mention the lack of truly viable options.[1] Think about the legal status of work trade, begging, and foraging on public or private land. People live by all of these strategies in the United States today, but they are hardly well regarded in our society, and practitioners rarely do without the market altogether. Too many of them live on the junk food they find dumpster diving.

The consequence is that any strategy of resilience has to take the market into account. But it has to include more. The good news is that agriculture can provide the surplus that makes integration into a larger economy possible. The bad news is that surplus production puts us at the mercy of whatever mechanism society has devised to extract that surplus from us, whether it's manorial dues or the Chicago Mercantile Exchange. The bad news is also that much of the natural resource base that traditional agriculturalists would have depended upon in crafting a livelihood is no longer available to farmers. Here we'll look at traditional subsistence strategies before turning back to our entanglement in the modern market.

## Tip-to-Tail: The Whole Hog of Subsistence

Fashionable chefs in my part of the world have been rediscovering the virtues of using the whole animal to produce fantastic meals. Never lost in Europe, the habit of using every available part of an animal has eroded in America (much like our topsoil) to a few prime parts cooked in a few predictable ways. The rest was considered scrap. Organ meats exported to poor cultures; hides rendered with the fat for animal feed, soapmaking, and the ever-dwindling lard market; head, bones, and the remaining tissue sent to the landfill. We've treated our landscapes in much the same way, cherry-picking the old-growth forest, clear-cutting the new, burning the "brush," and starting over, always looking for the fastest return on investment possible. Farmers, like butchers and diners, have long since lost the memory of how to use their farmscapes to the best advantage.

Traditional societies used the whole hog, and not just in butchering animals. The lands surrounding a settlement were precious resources for traditional agriculturalists; they were carefully husbanded and apportioned out among the members of the community on customary lines, often with the aim of rough equity. We will consider such "commons," their governance and management, in a later chapter. But strategies of stewardship to ensure continued access to resources such as firewood, forage, pasture, wild foods, fish, and game are found throughout the world's agricultural societies. To name just one, coppicing—the practice of cutting back shrubs and trees to the ground to promote vigorous regrowth—has been widely practiced, from the basket-making cultures of California to rural England. Coppiced willows provide straight willow wands for basket-making. Coppiced hardwoods were harvested for bundled sticks of "faggots" for firewood in wood-short late medieval Europe. Coppiced oaks provided poles and timbers for construction. Renewed in this way on a regular basis, coppiced trees tend to live longer and ensure a continuous supply of wood to users. Beyond woodlands management, traditional gatherers and agriculturalists carefully husbanded wild stands of berries, edible roots, herbs, and mushrooms. And many cultures were acutely aware of the multiple uses and stages of growth of valued plants, often having more terms for cultivars and wildcrafted plants than contemporary botanists.

These are just some of the practices that ensured continuous access to crucial resources for communities. We shall see more of them in the chapters to come. The important point here is that subsistence, and thus resilience, traditionally depended upon much more than cropping. Even on traditional American farms, subsistence included the family garden and what could be stored for winter consumption; milk animals, mainly for family use; poultry, pigs, and perhaps cattle; foraged berries and herbs; and the woodlot.

Farmers without such advantages were more exposed, more vulnerable, more stressed than others. American farmers who gave them up in the rush to farm fencerow-to-fencerow with modern machinery lost the farm—often literally, but figuratively, too, as they lost what makes a whole farm, a resilient farm, tick. Those uncounted other pursuits of traditional farmers can make the difference between the survival of the farm and its disappearance.

I once visited a Mexican village that seemed strangely empty. While talking to some of the men, I realized why. There were no animals. Though the village was surrounded by only marginal cropland, there was plenty of forage for animals. But when international coffee prices fell, depriving the men of necessary work in the coffee orchards a few hours away, they began seeking jobs far to the north, taking as much of their families with them as they could. Those left behind could not afford to keep animals. But for most Mexican peasants, as for traditional agriculturalists around the world, pigs, chickens, and the occasional goat or cow are major sources of savings, providing income in a medical emergency or to pay for shoes and books in the fall when the children go back to school. They also provide a bit of protein to go with a steady diet of tortillas and a few beans, and they are the ingredients in a wedding or saint's day feast. This village had given up all of that in the quest for reliable paid employment, and it was on the verge of disappearing altogether. The farm family that has reduced its farm to a house, equipment shed, and so many acres of cropland has migrated from the comforts of the traditional farm economy just as far as those Mexican villagers. And their farm is just as liable to be wiped from memory on the next economic downturn as was that village.

## The Cautious Farmer

Peasant farmers are notoriously cautious, shy of strangers, experts, and government officials, and unwilling to try something new, especially if proposed by any of the latter. Extension agents in the United States have spent decades training farmers to take their advice, and the educational system pushes the children of farmers, when they are foolish enough to persist in the idea that they want to farm, into agricultural schools where reliance upon experts is the first lesson of every class.

European colonial officials and, later, aid workers reported with despair the reluctance of the farmers they encountered to adopt the latest modern crop or tool. Techniques of persuasion have been marshaled against such stubbornness ranging from outright coercion to cash payments to peer pressure. Stalin starved the peasants of Ukraine into surrendering their crops, then herded the survivors into state farms and so-called collectives. Mexican peasants in the 1970s were given free fertilizer in exchange for a vote for the ruling party and the adoption of Green Revolution crops. In the 1980s, in international aid circles, farmer-to-farmer persuasion was preferred, paying early adopters to train laggards in the latest crop or fertilizer or technique.

The reluctance of traditional agriculturalists to adopt the experts' advice has been grounded in both long experience and their subsistence-first strategy. Farmers know from experience what the agricultural agents often don't know—that is, which crops and techniques and strategies have delivered a reliable living year after year after year. They sometimes claim that what they are doing they learned from their fathers or mothers, grandmothers or grandfathers, and that it is what their people have always done. Sometimes anthropologists or historians can show that this is not quite the case, that innovation does in fact take place—otherwise, how do we explain the enormous diversity of crops traditional agriculture developed long before plant breeders became certified experts? Nevertheless, the experience of farmers grounded in long practice often tells against expert-driven change.

Kevin Healy tells the story of his first experience as a Peace Corps volunteer in Bolivia, where he eventually returned to work for the Inter-American Foundation with the wisdom of hindsight guiding his relations with local food producers. Through dint of youthful enthusiasm, he had persuaded Andean peasants to adopt the latest creation of American potato breeding. All went well until the frosty day when the harvest began. With cries of consternation the peasants pulled one blackened, rotting potato out of the ground after another. Healy's story prefaces his own account of the arrogance of America's international agricultural aid efforts, which presumed to teach indigenous Andean peasants, whose ancestors had developed hundreds of varieties of potatoes, each with its own niche and uses, what sort of potato to grow for maximum yield and what sort of sheep should replace the cantankerous alpaca.[2]

Traditional agriculturalists with such a store of knowledge not only know more than their would-be advisers. They are reluctant to risk their subsistence with radical innovations. That's why the early adopters favored by the aid agencies are so valuable. Those farmers who are willing to take the risk first are often able to show others that it is worth taking. That they are not always right, that they are often the more prosperous farmers to start with and thus better able to take a risk, means that others still have reason for caution. Most peasants opt for what we are calling resilience, in other words, rather than profit maximization. And that choice is grounded in their commitment to their families and to farming for subsistence. Such a choice may frustrate advocates of market-savvy adaptation—and it might arguably need balance with a greater willingness to adopt new things—but it is a rational strategy for farmers committed to the long haul.

## Cultivating Diversity

The other side of the coin of traditional caution is the cultivation of diversity and of careful innovation that traditional farmers have practiced around the world. In place of constant, expert-driven innovation, subsistence-first producers rely on a diversified portfolio of crops, supplemented by livestock and the fruits of hunting and foraging, where those options remain. Andean potato producers have developed hundreds of varieties for the multiple ecological niches at their disposal, and each variety has specialized uses. Some can be freeze-dried in high-altitude sites. Some are grown for boiling. Some are medicinal. A single farmer may plant dozens of varieties. In India there may have been as many as ten thousand landraces of rice, with every village planting a dozen or more local varieties. Tolerance for drought, flooding, or saltwater intrusion; flavor differences; and medicinal uses all distinguished varieties. Corn in Mexico was similarly adapted, and Mexican farmers likewise did not grow just one variety but several. Even with the advent of high-yield commercial varieties, Mexican campesinos continue to cultivate their favored local varieties.

Diversity reigned across crops in traditional agriculture, and farmers relied on many crops throughout the year. Rice farmers in Southeast Asia treated yams, sweet potatoes, and cassava as "famine crops," available when rice failed. For the hill peoples these were staples, all the more attractive because they were not easily counted by the tax collector. But they also grew maize (after the sixteenth century), oats, sorghum, barley, and bananas. The forest gardens of Central America are diverse tangles of crops, as described by Edgar Anderson, which could include maize, more than one variety of beans and squash, chayote, banana, fig, plum, peach, avocado, agave, quince, and coffee in an area "about the size of a small city lot." The garden Anderson examined closely and mapped "was covered with a riotous growth so luxuriant and so apparently planless that any ordinary American or European visitor, accustomed to the puritanical primness of north European gardens, would have supposed . . . that it must be a deserted one. Yet when I went through it carefully I could find no plants which were not useful to the owner in one way or another."[3]

Since Edgar Anderson documented the diversity and high productivity of the tiny forest gardens of Guatemala, anthropologists, ethnobotanists, and a few economists have been noting the seeming chaos in the order of some traditional fields and the order in the chaos of others. The lack of the regular rows and neat subdivisions that appalled colonial officials and university-trained agronomists, it turns out, is part of highly productive systems that take advantage of synergies among plants, exploit ecological niches, and preserve and enhance genetic resources, among other virtues. The plots of highland swidden (aka slash and burn) agriculturalists in Southeast Asia and Central America, the fields of indigenous producers in northern Mexico, the American Southwest, the Andes, and much of Africa, and the scarcely recognizable gardens of Amazonia all present a picture that is vastly more complex and nuanced than the diversified agriculture of medieval Europe or lowland Asia on which modern, much more simplified methods have been based.

In swidden agriculture, as practiced throughout the world, woodlands are cleared and the woods burned, eliminating many pests and leaving valuable potash and phosphorus to enrich the soil. Stumps are rarely pulled, as they provide stability to the soil, particularly in rainy climates. And cultivation is minimal to nonexistent, many swiddeners using simple digging sticks to make holes for the seeds they plant. In the Southeast Asian highlands, where many peoples were refugees from oppressive rice paddy states in the lowlands, swiddeners planted a broad array of crops, taking advantage of differing dates of maturity and the potential for leaving root crops like cassava or sweet potatoes in the ground as insurance against times of flight, when government officials seeking taxes, conscripts, or slaves invaded the highlands.[4] In Amazonia, gardens scarcely noticeable to the casual visitor consist of dozens of plants transplanted or planted to fertile patches near a village.

Paul Richards, a conventionally trained agricultural scientist, found that West African polycultures could be more productive than the monocropped systems that development agencies have tried to foist on local farmers. Growing many different crops is, of course, one way to avoid risk, or rather to spread it and thus improve the chances of having food for the table. But it can also be highly productive. West African farmers, for example, practice intercropping, growing maize and sorghum with cowpeas and groundnuts, where the latter provide fertility for the former. Polyculture can also permit more plants per square foot than would be possible with a monocrop, and the crowding apparently provides mutual benefits among the partners.[5] And intercropping, Richards notes, is attractive for "the scope [it] offers for a range of combinations to match individual needs and preferences, local conditions, and changing circumstances within each season and from season to season."[6]

In the American Southwest and northern Mexico, researchers have found that indigenous farmers encourage wild cousins of cultivated maize and peppers in their plots, along with a wide variety of seeming "weeds." The farmers argue that wild teosinte invigorates their corn, even though tests show that the gene flow is small. Similarly, farmers in New Mexico permit wild peppers to grow up among their cultivated varieties to add heat to the cultivated chilies.[7] As Gary Paul Nabhan reports,

> Ask Pima farmers why they let volunteering plants persist among their intentional sowings, and they will give you a potpourri of answers. One plant may have edible

> greens, while another is host to edible insect larvae. Cacti and thorny shrubs emerging at the field margins can be shaped into a hedge that deters livestock from the plantings. Annuals may be allowed to shade the ground or certain seedlings for a while, but are then cut and used as mulch. . . . Trees like mesquite are left to shade tired workers from the desert heat, and their leaf litter is said to make the surrounding soil richer.

And, Nabhan says, in the most stable Native American villages, fields, hedges, ditches, and dooryards offer habitat to a rich variety of plants and animals, sometime endangered elsewhere in the same bioregion.[8]

In both Southeast Asia and Africa, New World cultivars were quickly integrated into traditional agriculture, making clear that innovation was not as difficult for village farmers as later Western extension agents found. The difference is that the extension agents and "development" specialists were pushing single varieties of commercially productive crops, like the Green Revolution corn, wheat, and rice. Traditional cultivators, on the other hand, were interested in subsistence and the contributions that the new crops could make to that end. And even where they wished to exploit market possibilities, they were careful to integrate the new market crops, like improved potatoes in the Andes, into diverse planting strategies.

West African farmers, Richards found, were using multiple techniques and readily adopted new crops as appropriate, as part of a broader pattern of experimentation and breeding for specific conditions. Rice farmers in Sierra Leone have bred their own varieties almost from the introduction of rice, and they carefully choose varieties for the soil type and moisture content of the soil.[9] James Scott concludes,

> Actual cultivators in West Africa and elsewhere should . . . [be] understood as lifelong experimenters conducting

> infield seasonal trials, the results of which they incorporated into their ever-evolving repertoire of practices. Inasmuch as these experimenters were and are surrounded by hundreds or thousands of other local experimenters with whom they share research findings and the knowledge of generations of earlier research as codified in folk wisdom, they could be said to have instant access to the popular equivalent of an impressive research library.[10]

The flip side of this observation, however, is also important: When moved out of the site of such experimentation, the knowledge, techniques, and even crops developed there can fail peasant and indigenous farmers on new terrain. A lot of ecological damage and farm failure has been associated with the forcible uprooting and transplanting of traditional agriculturalists by war, state planners, or economic migration, as the settlement of the arid West of the United States attests.

## Of Debt and the Market

Closely related to the widespread caution regarding expert-driven innovation are traditional attitudes toward debt. Debt is endemic in some peasant societies and a principal reason for poverty. Wherever farmers must bear significant start-up costs at the beginning of a season, they become targets for shrewd moneylenders eager to make a buck, and often considerably more, off their need. Whether that need is rental of a team of oxen to plow their field or a loan to hire the crew to prep the soil and seed the first plantings of the season, farmers turn to lenders for operational capital. As we shall see, some traditional practices eliminated that need, and with it the opportunities for wealthier individuals to set themselves up as intermediaries between the farmer and a crop. But where those systems of solidarity have broken down or are nonexistent, farmers come to depend upon the moneylender. In the unregulated markets of too many peasant societies, lenders take tremendous advantage of the situation, lending at usurious rates or demanding payment in the crop *a flor de la cosecha*, as they say in Mexico: at the moment of harvest, when it is worth the least. The moneylender will then store the crop until prices rise, in the worst of cases reselling the corn produced by his victim to the same farmer and hungry family at a premium price.

The elaborate institutions of many feudal societies were designed precisely to block this sort of practice, and moneylenders have had an evil reputation well beyond medieval Christian Europe and the ancient Hebrew world. Fair lending practices in the modern world take some of the edge off debt. Nevertheless, farmers traditionally have been as wary of debt as they have been dependent upon it. Again, agricultural extension agents and other government officials in this country have been at pains to erase this attitude, but it persists, particularly among older farmers. And with good reason. Agriculture is risky. Going into debt to carry on or expand adds the risk of losing the farm to the risk of losing the crop. Crop insurance was devised to lighten that risk, but it is not available to the really small farmer or the diversified farm, and it reduces but does not eliminate the risk. It does not make the payments on that brand-new combine when commodity prices collapse. Witness the continuing bankruptcy of American farmers.

More than caution is involved in another facet of traditional agriculturalists' engagement with the modern world. I'm referring here to the willingness of farmers to continue farming even when it doesn't make sense from the economist's point of view. In the 1970s and '80s, Mexican agricultural officials and left-wing academics engaged in a long-running debate over the future of the peasantry. Studies found that campesinos continued to grow corn even when they could earn more as laborers and buy their corn or tortillas in the marketplace. And campesinos were often aware of that fact. So why did they continue to farm? The answer was clearly mixed. Certainly they knew that paid employment was not secure, that it could take them far from home, that their families' subsistence depended upon a basic supply of that one staple, corn. But like the Maryland farmers I knew, they also loved farming and had little taste for the alternatives. And they loved their own varieties of corn, which many grew despite having little market demand. They clung to farming, in other words, because it was more than a job; it was a treasured way of life.

## Playing to the Market in Uncertain Times

Times are always uncertain for farmers, so the sort of caution discussed above will always be appropriate. What it means in translation is always the crucial question. Fortunately, a lot of young farmers have begun answering that question, picking carefully through the business advice offered by nonfarmers (and a few actual farmers). Here are some sample answers and some examples.

Small is beautiful. In fact, small scale is "one of traditional agriculture's basic strategies."[11] It may be true that industrial farming can only be viable above a certain increasingly large scale. But small-scale agriculture is tremendously productive. So productive, in fact, that economists and sociologists are constantly trying to figure out why the "inverse relationship" between size and productivity comes out that way in study after study. Can peasants really be better producers than agribusiness? The answer is most often yes. And that's whether we are talking about high-calorie crops like corn or wheat or "specialty crops" like fruit and vegetables. Small-scale agriculture benefits from all sorts of advantages—so many, in fact, that economists and sociologists are at war with one another over just which one is crucial.[12] Is it more labor? Clearly that's part of it. Is it greater attention to crops? Yes, again. Could it be better care for the soil? Undoubtedly. How about lower input costs? Of course. Is it mixed crop production? Integration of crop and livestock production? All these variables are relevant, and we'll find all of them deployed in traditional farming systems around the world.

Small farming, even micro-farming, can provide a modest but decent living to a farm family even under today's conditions. Eliot Coleman's acre and a half in Maine, Jean-Martin Fortier's similar Quebecois farm, Paul and Elizabeth Kaiser's three-acre farm in Sonoma County, California, and Ben Hartman's half-acre Goshen, Indiana, farm are all examples. And there are lots of less famous ones. These are farms that regularly gross more than $100,000 an acre through intensive, year-round cultivation. Compare that to the $400 gross revenue per acre for the average midsized farm. These micro-farms are mostly hand-cultivated, planted intensively both in the space provided and over time, and rely on direct marketing for most of their income.

Small farms like these do not require huge start-up costs, they produce much of their own fertility, they can minimize expenses and largely avoid debt, and they can provide much of a family's subsistence even when markets falter. Jean-Martin Fortier offers this advice: "Start small and then gradually add on. Focus on specialized processes that you enjoy doing."[13] Eliot Coleman's *New Organic Grower* makes the same point and shows how to appraise, and occasionally invent, the tools that will best serve the farmer.[14] Ben Hartman and his wife, Rachel Hershberger, built four hoophouses, borrowing for another only when the first had paid itself off. That kind of cautious expansion is part of a larger pattern of carefully assessing what works and what does not, what makes for wasted time and effort and what can enhance efficiency. Hartman calls such thinking "lean farming." Lean farming led Ben and Rachel from three acres and a large tractor to a half acre and mainly hand tools. It also led to the sort of record keeping that would please any farm economist, but put that effort into the service of providing more and more of what their customers wanted, not courting a bank loan officer.[15]

Paul and Elizabeth Kaiser also gave up their tractor and shrank their farm, opting to build soil with heavy doses of compost and a refusal to till. The Kaisers never leave a bed bare, maintaining fertility through applications of compost, leaving crop plant roots in the ground after harvest, and careful rotations—as many as eight a year. And they practice the sort of polyculture we have seen in traditional farming around the world. They sell through a year-round CSA, farmers markets, and direct to restaurants, earning enough to support their family and four full-time and year-round workers.[16]

Perrine and Charles Hervé-Gruyer came to intensive gardening by way of permaculture. Having planted and landscaped an elegant orchard and forest garden, they began market gardening under the influence of John Jeavons and Eliot Coleman. Jeavons and Coleman draw on the heritage of nineteenth-century Parisian market gardeners, and both have developed methods of intensive planting that exploit all the potential of a small plot of land. Permaculture principles helped Perrine and Charles build their soil, while Jeavons's spacing and Coleman's six-row precision seeder and other tools provided the techniques. Perrine and Charles improved on both, adopting such tricks of the Parisian gardeners as planting slow-germinating carrots and fast-growing radishes in the same row, so that by the time the radishes were harvested, the carrots would be leafing out.[17]

These are all market gardens and rely principally on vegetable sales for their income. More diverse integrated farms demand more space but can provide more resilience, a wider range of production, and often more of the ecosystem services that farms once depended upon. Joel Salatin's Polyface Farms comprises 100 acres of pasture and hay field, plus 450 acres of woods, sustaining hundreds of animals, all primarily pasture-fed. The Hervé-Gruyers' Bec Hellouin Farm in Normandy, France, profiled in *Miraculous Abundance*, concentrates hundreds of fruit trees and berry bushes, a half-acre intensive market garden, ponds and watercourses, and space for small livestock in just two and a half acres. The Hervé-Gruyers envisage a farming system based on micro-farms like their own supporting small-scale grain and cattle production, a food forest, and nature reserve, as well as the laborers, cooks, and craftspeople such a community would need to thrive.

Salatin and the Hervé-Gruyers represent the sort of "middle farm" that Wendell Berry and his daughter Mary are trying to revive through the work of the Berry Center. But both of these examples are what the Berrys call "entrepreneurial farmers," farmers, that is, with the sort of flair for marketing that many of us lack. What about those farmers who just want to farm? There are examples here, too, and some promising innovations.

One of the best-developed of these is holistic management, applied especially to livestock operations. Pioneered by Allan Savory and now advanced by Holistic Management Institute International and the Savory Institute, techniques of mob grazing and careful pasture management have been adopted by hundreds of cattle and sheep producers around the United States and recently celebrated for their ecological benefits. Savory's methods were based on observation of wild herd animals in southern Africa. Dismayed at the ecological destruction cattle seemed to bring with them, Savory began to analyze what was different about the grazing habits of wild herbivores in the region. Still harassed by large predators, these animals tended to bunch together (hence "mob grazing") and consume available nutrients before moving on. The churning of the soil, even in wetlands and riparian areas, Savory found, brought new seeds to the surface and renewed growth. So did the grazing, which stimulated growth of grasses and forbs adapted to grazing. Moving the herd along systematically optimizes this regrowth, which also has the effect of stimulating root growth, particularly in pasture grasses, sequestering carbon in the process. Pastures recover and flourish under such management, increasing holding capacity, unlike the degraded pastures characteristic of so much extensive grazing around the world. And with higher holding capacity and healthier pastures come healthier, more valuable animals.

Grain farmers in the upper Plains are using polycultures of cover crops and diverse crop rotations, sometimes coupled with grazing, to improve soil quality and deal with the increasingly erratic climate conditions they face.[18] Intercropping and pasture-cropping are also adding to the tools that grain farmers have at their disposal. With intercropping farmers can grow two or more crops at once, typically a grain and a legume, like the West African farmers. Yields are lower for each but higher per acre with two crops in one space over the same season. Pasture-cropping has been developed especially in Australia, where sheep rancher Colin Seis has dramatically increased yields on vulnerable land by sowing a grain crop after pasture is spent and harvesting just as the pasture grasses start to regenerate.[19]

Permaculture has a lot of cachet but few outstanding examples of productive farms or farming systems on the scale that could support a family or feed a community. That doesn't mean there couldn't be such farms, as the examples from Guatemalan and West African traditional agriculture show. The polyculture of Bec Hellouin farm appears to be highly productive, though the farm's principal income is from intensively planted market gardens. Fruit trees, ponds, and livestock contribute to the resilience and self-sufficiency of this farm, but they aren't the main actors in the marketplace. Mark Shepard has developed a model he calls "restoration agriculture" that exploits the advantages of tree crops and the midsized, integrated farming operation familiar to the West. The foundation of Shepard's model and the chief calorie crop is the chestnut, planted in imitation of the oak savanna that once characterized much of the North American landscape. Widely spaced, chestnuts are a reliable nut crop that can provide feed for hogs as well as humans, while leaving room for a variety of other crops. In the understory, for example, and along waterways, hazelnuts produce a second calorie crop, while berries of all sorts bring high returns in the marketplace. The alleys are mowed for hay or pasture-cropped, then followed by a succession of livestock utilizing mob grazing techniques and portable fencing to keep the livestock from the trees and berries, starting with cattle, moving on to sheep and goats, then to pigs and poultry.[20]

## Strategies for the Long Haul

These are justly celebrated examples of how to confront the market we face today with invention, sustainable practices, and resilience. But the prospects that current market conditions will govern our lives ten or twenty years from now are low. The resilience on which subsistence must be based means being in a position to adapt as petroleum supplies falter, markets go erratic, or financial crisis makes debt and further borrowing unsustainable. In the meantime, we have to get a living and do so in a way that develops the tools and skills needed for an uncertain future. A Saudi prince may advocate keeping the oil in the ground to provide the basis for the plastics of the future, but we are not going to do that, and who wants more plastic anyway? Until oil depletion makes both transportation and plastic outrageously expensive, however, we will continue to use both.

My own micro-farm uses greenhouse plastic, synthetic row cover, shade cloth, and landscape cloth, all petroleum-based. That's how we keep up the year-round production that makes ours a viable market garden, not to mention providing vegetables for ourselves and our community in the winter months. Significant strides in producing non-petroleum-based agricultural plastics are not in sight. So we provide for ourselves and our neighbors now with strategies that will be finished when the Age of Oil is over, not so many years down the line. Plastic keeps us on the land and gives us time to develop new strategies, but we had better get to it.

The same is true of transportation. In my area we have recently begun to fill in the "missing piece" in the puzzle of developing a local agriculture, a food hub that can move food from one part of our scattered rural county to another. All very good, but our oil footprint, however small relative to the well-traveled, fifteen-hundred-mile head of cabbage, is still an oil footprint. As far as I can see, the prospects for an electric truck fleet to keep that food moving are dim, Elon Musk notwithstanding. And those trucks, solar- or wind-powered, still have to move on roads that, up to now, are maintained with major fossil fuel inputs, from diesel fuel to asphalt to concrete. The local foods of a resilient future are likely to be *very* local; our task is to build the capacity to provide most of the calories for ourselves and our neighbors before the crutch of the Age of Oil crumples. In the meantime, though, we need that food hub to have the ability to produce and sell in a world still dominated by Big Ag and bigger supermarkets. Because if all the local farmers have quit because they can't make a living today, there will be no resilient food economy to feed us in the not-so-distant future.

Resilience, then, has at least two big parts. One is to build the capacity to feed ourselves and our communities now. And that depends on adopting the best strategies for keeping ourselves afloat and farming for the near future. We've described some successful examples of doing that above. The other is to develop the tools and techniques for a future when many of the resources we rely on now either will not be available or will be financially out of reach. The subsistence strategies of the past give us lots of hints about how to do the second of these.

One of the key lessons of the persistent agricultural systems of the past is a dependence on diversity. Our range of products, our production strategies, our ability to produce for year-round consumption are all key to our subsistence and thus to our survival as farmers. When our primary market fails us, when the cost of gas and propane skyrocket, when groceries are just too expensive, farms that have livestock for family consumption, firewood readily available, plenty of vegetables, and calorie crops like potatoes, flint and flour corn, dry beans, and winter squash will be better able to weather the storm and face the future.

Being able to muster a variety of growing techniques, from dry farming to season extension, and growing crops with differing requirements will increase our ability to face changing weather patterns. Biological diversity is thus key to our future. Homesteader and plant breeder Carol Deppe has stressed the need to develop new varieties on-farm as we face new diseases and climate change.[21] Many farmers already save seeds and have learned or are learning the rudiments of plant breeding. Even in today's market, as another homesteader, Will Bonsall, says, if we keep buying seed we remain a slave to the marketplace and encourage breeders to produce hybrids and trademarked varieties that keep us captive.[22] Worse, we will see the genetic diversity on which adaptation to *our* climate and *our* conditions depends continue to be eroded. And developing a diversity of varieties of even a single crop increases resilience. Like the Mountain Pima farmers in Mexico who hold in reserve two or more maize varieties specially adapted to shorter growing seasons or a cold, wet start, we need to plan for adaptability.[23]

We can also begin to imitate the maize and chili producers of Mexico and the Southwest and encourage wild diversity within our farms. Farmer Bob Cannard is notorious for tolerating "weeds" (or at least certain weeds) among the crops on his Green String Farm in Petaluma, California, encouraging biological diversity and exploiting the talent of some weeds for providing nutrients for the soil. Orchardist Michael Phillips makes the same case for the orchard floor.[24] For the more fastidious, carefully selected and planted hedgerows and perennial flowers can host beneficial insects and small predators that enrich the garden, orchard, or field.

Diversity also applies to the resources available on the larger farm. Can we produce our own fertility with resources on the farm? Doing so is not just an answer to long-term sustainability but an ingredient to a lower-cost, lean farming operation today. Bob Cannard, like John Jeavons, believes a portion of cropland needs to be devoted to producing high-biomass crops for composting. The

Hervé-Gruyers utilize crop residues and locally produced hay for the same purposes. Will Bonsall harvests grasses from his meadows, leaf litter from his forest floor, and ramial wood chips from tree trimmings for compost and mulch on his Maine farm. Animal manure and bedding have traditionally played the same role, though Bonsall insists that the more direct and efficient method is to use the plants that the animals would otherwise consume, both for compost and mulch and as green manure in the form of cover crops. Anyone who has built a compost pile or turned in green manures may disagree with the efficiency evaluation, but animal manure must be composted or turned in, too.

Then there's the woodlot—where *will* your heat come from when the fossil fuel tank is empty? Where will the building lumber come from when the long-haul trucks stop running? It may be that solar energy can be harnessed for some time to come with available technology and that we can all cook with electric stoves and heat our houses with electric heaters, as Richard Heinberg and David Findley have recently argued.[25] But having solar electricity means acquiring the ability now, while we still can, because photovoltaics depend upon rare metals mined in conflict zones like the Congo and combined in hermetically sealed and air-conditioned manufacturing facilities that may be hard to maintain in the not-so-distant future. The resilient farmer will hedge her bets, acquiring off-grid solar capacity now, if possible, *and* making sure that alternative sources for cooking food and warming the house are in place. We have to start, of course, with conservation. Similarly, we might build new, energy-efficient homes with heat pumps and the like, supposing we can afford those extravagant measures. But we had better build them now, while the technology is cheap. In the long haul, many of us will still depend on that woodlot, and learning to nurture it, coppice it, and use it wisely are skills we will need to build now.

The same sorts of considerations affect much of the technology farmers are currently using, from record keeping to transportation to tractors. John Michael Greer suggests that we return to the concept of appropriate technology, and that many of the technologies we will need for resilience are what he calls "trailing-edge technologies," those that were once perfectly serviceable but were bypassed by the market. Examples abound, from ham radio to treadle sewing machines.[26] It is all very well in the present market economy to tout the advantages to the entrepreneurial farmer of cutting-edge technologies like cell phone apps to track planting, harvesting, labor time, food safety compliance, or sales. But the enormous infrastructure of the internet and cell phone communications is a likely casualty of any serious collapse of the fossil fuel economy.[27] That hasn't happened yet, though net neutrality is gone, and we have to make a living as efficiently as we are able today. But it would be well to have the working knowledge of math and manual record keeping and accounting that electronic apps are replacing. And what about transportation, traction, and all that plastic in our farming system in the future? What will replace it all?

The most resilient farms of the near future will be the least dependent on petroleum-based technologies, but that raises the question of what technologies we should be preserving. We can name some trailing-edge technologies as a first step in beginning to think about how we might adopt them. Wendell Berry has long made the case that animal traction on the farm reached its technological height sometime late in the nineteenth century and that it remains perfectly suited to small-scale farming and forestry today. Draft animals can also supply the more limited transportation needs of the future. But we will have to have the animals, the tack, and the knowledge that were all abandoned seventy or eighty years ago in this country. Glass once played the role of plastic in the intensive market gardens of Paris, the Netherlands, and England. There's plenty of discarded single-pane and even double-pane glass around today, thanks to ever-tightening building codes, and today might be a good time to start rebuilding those propagation and hoophouses with more durable glass. It won't last forever, but it will carry us a good bit longer than our five-year greenhouse plastics.

And today is as good a time as any to begin learning traditional techniques of slaughter, butchering, preserving, canning, drying, milling, baking, and long-term storage of meat and crops. These aren't just the affectations of foodies nostalgic for an almost forgotten and romanticized past. They are practical skills that we need to learn now, because we will be needing them soon enough and they can serve us well right now. Just as we need resilient farms and farmers, we need to cultivate the specialists in all these old-fashioned technologies. Luckily, most rural communities around the country still have trained butchers who can also slaughter and dress a steer or hog. The butchers were put out of work by the big, consolidated meatpacking industry but have found work in their own custom butcher shops serving 4-H families, hunters, and farmers who prefer their own meat over the dreck in the supermarket.

Lots of knowledge about preservation of fruit and vegetables is still available and in use. The home canners are carrying on old family traditions in the name of economy or in search of a few dollars at farmers markets and other small outlets. These are people we need to support and learn from, the sooner the better. Meanwhile, the foodies have developed traditional skills never widespread among the larger population, from all sorts of fermentation to brewing kombucha. Many of these skills are still employed in the marketplace and not the marginal avocation depicted by ag advisers who insist farmers keep our eye on the bottom line. They are, in fact, the key to *having* a bottom line that keeps us well-fed in good times and bad. The following chapters will take up some of these strategies for resilience and spell out in more detail the examples from the past that can inform our future.

Chapter 4

# Land for the Tiller

La tierra volverá a quienes la trabajan
*Let the land return to those who work it*
—Zapatista slogan from the
Mexican Revolution, 1912–17

Good farming entails stewardship of the soil, but that depends crucially upon secure access to land. In this chapter we consider access, that is, how cultures, including ours, have provided land to farmers. In the next chapter we look at how farmers have managed that land for the long haul. Access and use are inextricably intertwined in the history of agriculture. Access to land comes first in farmers' priorities, but even where land suitable for farming is scarce, human cultures have found ways to transform marginal land for production. And abundant land can encourage misuse and abuse of the land and natural resources, as it has in the settlement of the United States. Ownership patterns, moreover, can make even ample resources scarce for most farmers.

In settled cultures, all sorts of factors can make access to land problematic, from land law to urban sprawl, from speculation to demand for high-value crops. Resilient farmers and the communities that would sustain them have to navigate those challenges and find ways to ensure continued access to land for farming. Contemporary Amish farmers carefully save money to acquire land so their children may farm on a scale that can provide for the families of the next generation. They are so successful at it that they have recovered expensive land around traditional Amish settlements in Lancaster County, Pennsylvania, where suburban sprawl has long threatened farming. But such cases are rare in the United States outside the Amish communities. Farmland is disappearing at alarming rates due to sprawl, highway development, and speculation. The American Farmland Trusts estimates that over forty acres of farm- and ranchland are lost to other uses every hour. Most farming today utilizes rented land; the USDA reports that over half of cropland is rented. But rental too often provides insecure tenure and, without security of tenure, few incentives to build the soil or the infrastructure that farmers need. The majority of rented farmland, moreover, is in the hands of nonfarmers, and that percentage is growing.

Everywhere today access to land has become the compelling issue for those who want to farm. In the United States it is estimated that 70 percent of farmland is in the hands of farmers age sixty-five and older. That four hundred million acres will change hands soon, but how? And to whom? In England the 1947 Town and Country Planning Act was meant to protect the countryside from the sort of wanton urbanization that was already overtaking the peripheries of the cities. Strict planning guidelines, combined with land concentration for industrial farming, have put much of the countryside off-limits to rural residential housing and made it next to impossible to establish a small farm. The law has had the further perverse effect of promoting a speculator's market for existing farmhouses and village house sites, catering to those urbanites who can afford a second home in the country at the expense of the rural working class.[1]

Though most of us think that land concentration and high prices in the United States have been the result of impersonal "market forces," government policy has always played a large role. The Populist movement, which we will look at in chapter 10, revolved around the lack of credit for farmers and farm families in the deflationary, tight-money post–Civil War economy. In the aftermath of Populism's defeat, rates of tenancy rose as more and more farmers lost their land to mortgage holders. The rates rose from 28 percent nationwide in 1880 to 38 percent in 1910. Even in Iowa by 1935, 49 percent of farms were tenant-operated. The federal farm credit that was legislated following creation of the Federal Reserve in 1913 went first and foremost to the largest farmers and was then extended to what remained of middle-sized farmers, excluding those most in need of protection and making possible the acquisition of their land by wealthier landowners. As Lawrence Goodwyn, historian of the Populist movement, puts it, "In essence, 'agribusiness' came into existence before it even had the opportunity to prove or disprove its 'efficiency.' In many ways, land centralization in American agriculture was a decades-long product of farm credit policies acceptable to the American banking community."[2]

Even where market forces played a large role, governments had a hand in pricing the small farmer out of business. In California large concentrations of ownership, starting with the Spanish land grants, have always placed a premium on prime farmland, driving prices up. The California Project, intended by Congress to provide water to small farmers, quickly became the plaything of the largest farmers as prices for irrigated land soared. Suburban sprawl, and the failure of county zoning laws to stop it, has been adding to the pressure on farmland at increasing rates. Economists justify the practice as an example of the market determining the most economic use of the land. By the economists' logic land will only be preserved for farming when the per-square-foot value of production reaches that of the shopping mall. Few economists go so far, but Stephen Blank, of the University of California, suggests that we should give up food production in this country altogether and simply import our food from countries with lower land rents.

One result of the pressure on land prices has been increasing recourse to higher and higher value crops, even where smallholdings still prevail.[3] Parts of the Central Valley transitioned in the 1920s from wheat to fruit trees. Later, organic production began to compete heavily with conventional fruit and vegetable farms on fertile coastal plains and elsewhere. On California's North Coast marijuana now competes with wine grapes as the crop best able to sustain a small farm. But initial investment costs for both crops suit the well-connected and already wealthy, not aspiring small farmers; and the legalization of marijuana, with expensive permitting and regulatory requirements, is putting new pressure on already high land prices and driving smaller growers out of business. Meanwhile, in upstate New York big wine companies from Italy and California are buying up land to exploit new opportunities as global warming pushes wine culture north. And those are just the farming interests at work. Pension funds, hedge funds, and institutional investors of all sorts are buying land as prices rise and speculative investment appears worthwhile.

All over the world today, in Ukraine, West Africa, Brazil, and Southeast Asia, huge multinational corporate interests, some of them backed by governments like China, others by corporate charities like the Gates Foundation, are buying or leasing swaths of farmland in the hope of providing for the anticipated food and fiber shortages of the twenty-first century.[4] Small growers are being displaced to produce fresh flowers and vegetables for developed country markets, coconuts or coffee for gourmet consumers abroad, biofuels and animal feed for national and multinational corporations, and grain for the processed food industries of the world. Peasant owners, when they are not simply driven off with the connivance of corrupt local officials, are expected to welcome the slim dividends from sales or lease agreements and find alternative means of employment ("jobs") in an agricultural economy that no longer operates at the human scale.

In fact, access to land has always proven to be the crucial underlying dynamic of farming throughout the world. The beginnings of the enclosure movement in fourteenth-century England displaced peasants growing grain and produce for local consumption with sheep whose wool was destined for markets in the Netherlands and Italy. The ferocious privatization of English landholding in the eighteenth and nineteenth centuries, when over seven million acres passed into the exclusive jurisdiction of private owners, left a countryside that today supports only a third of the population of the contemporary French countryside.[5] Mass migrations to the white settler colonies of North America, Australia, New Zealand, and southern Africa were driven by a land-hungry European peasantry experiencing a population boom that was unsustainable on the available landbase. And peasant demands for more just systems of landholding fueled twentieth-century revolutions, from Mexico to China.

How a people provides land to farmers, and what sorts of demands farmers are forced to put upon that land, have shaped farming systems everywhere. The revival of small-scale agriculture suited to replace the failed agricultural system of the twentieth century will depend crucially on our ability to ensure access to land—and not just cropland—for the farmers of the twenty-first century.

## Private Property: A Historical Anomaly

Every farmer and aspiring farmer I know wants a piece of land that he or she "owns," that is, has a long-running, exclusive legal claim to. What few add in celebrating the merits of ownership are the risks of foreclosure should it become impossible to pay the mortgage, or of seizure through the heavy hand of eminent domain. Together these make up the rudiments of fee-simple ownership—exclusive right to use and dispose of property, accompanied by long-running debt (in most cases) and hedged by the claims of the state. This is the package we have come to regard as the ideal from the perspective of secure access to land.

This package of gains and risks also comes freighted with long-term consequences for farming that have not proven beneficial, some of which were hinted at above. It is particularly perilous for a renascent and resilient agriculture because of the speculative pressures that fee-simple ownership presents to any landowner. Even among farming families, pressures to sell rather than preserve the land in farming for family members who want to farm are often strong as the older generation retires, because brothers, sisters, and children who do not care to carry on the family tradition nevertheless want their piece of the pie. And the real estate market, whether driven by pure speculation, development pressures, or boutique crops, provides the opportunity for a family to cash out of such competing and conflicting interests to the loss of their aspiring farmers.

Private property has another perverse effect. It has virtually abolished the commons upon which more resilient farming systems depended for a long list of nonfarmed goods. Those commons typically included pasture, forest, wetlands, and even irrigation resources. From the commons farming societies gained forage for livestock, wild game, herbs and wild edibles, firewood and lumber, resources for weaving, clay for pottery, and much more. And the commons functioned, too, as an ecological reserve and buffer against too rapacious a human exploitation of nature. For contrary to popular supposition, historical commons were most often well managed. Some, indeed, have survived for more than a thousand years in much their present form.[6] The so-called tragedy of the commons is a bad metaphor founded upon a myth, as its author, Garrett Hardin, was eventually forced to recognize.[7] We'll look more closely at the commons and what they represent for a resilient farming culture in chapter 7.

The notion that land can be a speculative commodity, and one that may be disposed of wholly at the discretion of the owner, is a historical novelty, argues historian Andro Linklater, nurtured in the peculiar history of English struggles between crown and gentry reaching back to the thirteenth century. For most of human history, land has been widely shared (if at times tyrannically governed by a "lord" of some sort) and apportioned according to custom, and often need, to farming families. Security of tenure was achieved through custom and participation in a community in most societies, though subject to revocation in those rarer cases where states claimed sovereignty over the resources of the communities that supported them.[8]

In England in the thirteenth century, as in traditional societies around the globe up to our own day, access to land was part of a complex web of mutual obligations. If nobles had the right to command the labor or the produce of peasants, they also had the obligation to maintain them on the land and to provide sufficient cropland and access to commons to ensure their livelihood. Abuses were many, particularly because the noble was himself judge and jury in the manor court where disputes were adjudicated. For that reason, and as a part of a struggle for power between king and nobility, a series of English kings had established the rights of tenants, including the right to a hearing in the king's court. With the Magna Carta a cabal of noblemen turned the tables on a weakened King John, establishing the absolute right of an owner of land to trial by jury under laws duly passed by Parliament.

The eventual consequences were revolutionary. "In sharp contrast to Roman law," writes Linklater, "and for that matter to the civil obligations of landowners under Chinese and Islamic custom, the version that evolved under English common law had no counterbalance. Far from subjecting rights of individual ownership to those of social obligation, throughout the sixteenth century the law heaped civil liberties, political power, and legal protection upon the freeholder at the expense of everyone else."[9] The immediate result was that not only tenants, but everyone without landed property, were deprived of voice and authority in the emerging English constitution. Two waves of enclosures—privatization of common land and even of village sites—transformed English society and the countryside.

The first wave of enclosure culminated in Henry VIII's expropriation of monastic properties and their sale or gift to aristocrats and court functionaries. Ironically and fatefully the collapse of wool prices that followed forced many of those indebted large landowners to sell, and the beneficiaries were merchants and better-off peasants. Suddenly the liberties of the English property holder were extended to another class or classes: the gentry, the yeoman farmer. One notable exception was women. As the rights of property holders were codified, the law of couverture of 1542 abolished the right of a wife to a part in family property, something that had been guaranteed by Roman law and feudal practice. With the law of couverture, everything a woman owned would be transferred to her husband's name, a stipulation that persisted in English and American law into the nineteenth century. Linklater quotes the famous English legal theorist William Blackstone, who noted with some alarm that couverture meant that "the very being or legal existence of the woman is suspended during the marriage," and for purposes of inheritance even after her partner's death.[10]

Commons persisted for many of the villages affected following the first enclosures, even with private property in arable land. It took a second great wave of enclosures, starting in the seventeenth century and extending into the first decades of the nineteenth, to erase most of the rights to common lands that English peasants had enjoyed. And in the process it also tore many of these peasants from the land itself, creating the basis for the English working class and more than a century of conflict that only barely escaped erupting into open warfare. But for those who gained land in the process or migrated to the New World in search of it, widespread property ownership became identified with political freedom.

This is the version of property rights that established itself in the white settler colonies of the English Empire, and it was the common presupposition of the Founding Fathers of the United States. More sensitive to social injustice than many of his elite contemporaries, Thomas Jefferson deplored the effects of private property in Europe, inveighing against the poverty induced by gross inequalities in property ownership: "The consequences of this enormous inequality producing so much misery to the bulk of mankind, legislators cannot invent too many devices for subdividing property."[11] That may explain Jefferson's substitution of the word *happiness* for *property* in the list of rights derived from the work of John Locke that famously undergirds the Declaration of Independence. His proposed solution was what later would be called land reform. Jefferson at first proposed that land in America be distributed as leaseholds, not private property, in order to ensure continued equality of ownership. In the end, however, Jefferson himself was the author of the most far-reaching scheme ever devised for managing, and privatizing, the new lands of the United States acquired by war and purchase. The lands between the Appalachians and the Mississippi, and later the Louisiana Purchase, would be surveyed and sold in blocks.

Everywhere that private property spread, there, too, banking flourished, according to Linklater, because privately held land provided the necessary collateral for loans and often depended upon loans, both for purchase and for improvement. That connection also laid the groundwork for speculation, as those with money or access to credit grabbed land in anticipation of later demand. That was the fate of much of the land that Jefferson helped prepare for sale. Speculators were the frontiersmen of landownership, buying up whole sections with the intention to sell at a high price to the settlers who followed. The Homestead Act of 1862 proposed to stem that practice by making limited parcels of land free to those who "improved" it. But even after the Civil War enormous tracts of public land were up for cash sale with no restrictions on the size of holdings, and the link between fee-simple landownership and speculation remains in place today.[12]

Private ownership of land in one form or another, of course, has been found throughout human history and throughout the world. And it has often led to disastrous consequences for farmers.

Sumerian records show that poorer farmers could lose their land, their tools, and even the freedom of their wives, children, and ultimately themselves to creditors, who would make slaves of debtors. Reflections of these patterns can be found in the biblical prophets, who inveigh against the rich who "lay parcel to parcel" and enslave the poor. The prophets call for a "Jubilee year" of forgiveness of debts and release of captives and slaves, a practice first promulgated by Mesopotamian kings and dating back at least two millennia.[13]

Jefferson's solution to the problem of access to land (intended to secure the West in American hands) fueled speculation and debt, but it had another unintended consequence: The grid-line survey that he promoted created property in neat blocks, sections, and quarters. Resources like streams, forests, and wetlands that would have been commons under traditional practice were fragmented and incorporated into discrete chunks of private property with all use governed by ownership. As farmers fenced their land against neighboring livestock, these commons were further alienated from use by any but the tenant on the land. In the West, where conflicts between fencing farmers and free-range cattlemen were fierce and often deadly, private property became off-limits to all but owners. The English and European practice of allowing hikers access across open fields and through private woods became unthinkable, ownership enforced at the point of a gun.

The English system of private ownership spread around the world with colonization and on a wave of prejudice that said that older systems of land tenure were inherently unproductive. In eighteenth- and nineteenth-century England, a new wave of enclosures was justified on the grounds that "improving landlords" would make better use of the land than peasant farmers. Even then observers could show that smallholdings were actually more productive than the newly consolidated lands, but that did not stop Parliament from approving new enclosures. Some four thousand Acts of Parliament enabled the practice; around seven million acres, or one-sixth the area of England, were privatized. And whatever the judgment on the productive advantages of privatization, the process, as Simon Fairlie puts it, "was downright theft. Millions of people had customary and legal access to lands, and the basis of an independent livelihood was snatched away from them" by enclosure.[14]

The same process was emulated around the world under other legal systems, and everywhere the results have been much the same. Wherever Liberals won power in nineteenth-century Latin America, they moved to privatize both church and communal lands; wherever they succeeded they laid the basis for land concentration and, eventually, revolt and revolution.[15] The British Raj in nineteenth-century India turned traditional rulers into aristocratic landlords, setting in motion the dispossession and impoverishment of millions. In recent years the World Bank has been as eager as nineteenth-century British colonial authorities to dissolve communal landholding systems in Africa and elsewhere and replace them with fee-simple ownership. And in 1995 a neoliberal government in Mexico reversed the land reform that had helped create the Mexican state, opening the way for Mexican holders of *ejido* (*eh-HEE-tho*) property—the village-level vehicle for communal landownership in Mexico—to sell off the land that had been guaranteed them and their communities in perpetuity.

Everywhere the argument in favor of enclosure, a "free market in land," and consequent speculation has hinged on economic gain. Either the "best" (in other words, most profitable) use lay outside farming altogether, for example in building a housing tract, or increased scale of operations made the consolidation of landholdings a welcome outcome. Unfortunately for the argument, the evidence for economies of scale in agriculture is depressingly consistent: There are none. Not at the level of the giant farms that dominate the world today. And scarcely at the level of the typical medium farm in this country. In 1967 a USDA economist reviewed 138 studies on the production costs of different-sized farms of all sorts and found that increases in efficiency were nonexistent or even negative above a small- to midsized farm.[16] Countless recent studies confirm those results. As George Monbiot puts it in a summary of studies from around the world, "There is an inverse relationship between the size of farms and the amount of crops they produce per hectare. The smaller they are, the greater the yield."[17]

The popular press often confuses the issue by asking a different question: Can organic agriculture feed the world? It turns out that larger-scale organic agriculture is generally less productive and more expensive than conventional agriculture. It may be better for the earth (or not) and better for you (or not); but it doesn't have the magic bullets that larger-scale conventional agriculture deploys in the form of chemical inputs. As farmer and permaculture advocate Chris Newman provocatively puts it, "Organic farming is little more than conventional farming with all the tools taken away. It's a well-intended but insane way to farm, and it will kill us all if we decide this is the way to 'fix' agriculture."[18]

By contrast, as we saw in the last chapter, small-scale agriculture can be tremendously productive. While Marxists and neoclassical economists see small-scale production as a relic of the past, inevitably doomed to fail (however productive) English farmer and social scientist Chris Smaje insists history may be pressing us in the opposite direction, to "re-peasantization."[19] Certainly small scale is the approach many of us are taking today in response to the failures of industrial farming and the limited access to land that we face. And as we struggle to extend access to land to more new farmers and shape the farming societies of the future, we would do well to take into account older systems of land tenure.

## The Relevant Past

What did access to land look like before the private property revolution in the West? The answer varied enormously across societies, and some of the alternatives are not attractive. But there are countless examples of institutions and practices that have provided widespread access to land and natural resources on reasonably fair and sustainable terms. One practice that was *not* common, at least among agriculturalists, was communal use of the land, communal farming. Shared pasture and woodlands have often been documented. But traditional agriculturalists have generally not worked the land in common unless serving as agricultural laborers on some overlord's estate. Otherwise, traditional arrangements have typically allotted land permanently or temporarily to one family or farmer for cultivation. The conviction of both early-twentieth-century American farm advisers and Joseph Stalin that higher efficiencies could be obtained by collectivizing work on the farm was not shared by traditional agriculturalists, who were targets for liquidation or displacement by both parties just for that reason.[20]

Individual *use* of the land, however, rarely has meant individual *ownership* of the land. And even where traditional agriculturalists "owned" the land in some sense, ownership came with obligations and stipulations that the modern system simply and brutally abolished. Those obligations ranged from shared decision making about cropping and the annual distribution of plots to opening harvested cropland to communal gleaning and grazing to providing for the subsistence of poor farmers, as well as their widows and orphans.

The open field system found in England and much of Europe distributed land in strips through annual meetings in the manor court or village green. In the best of these arrangements, peasants would share among themselves strips of better and worse land, and together practice a two-year or three-year crop rotation. The village ox team would plow each family's allotment in turn, relieving poorer peasants of the expense of maintaining a team. The oxen would be fed off the commons and the hay harvested from the commons. After the harvest all the strips would be open to gleaning by the poorest of the village, then to communal grazing by whatever livestock peasants possessed. The downside of the system was that individuals were not free to choose their style of farming but had to conform to the general outlines of the community's farming system. The upside was the use of that team of oxen, secure access to land, and a common sense of responsibility for the sustainability of the soil.

Variations on the open field system were (and still are) found around the world. In France, Simon Fairlie reports, open fields were considered the most productive systems in the early modern era, and they go back to the earliest settlement of the French countryside. They persisted into the nineteenth century despite being deprived of legal standing by modernizing legislators, simply because they worked so well.[21] In Germany and Eastern Europe, they were the basis for the colonization of new lands from the Middle Ages onward.[22] In the Tigray Region of Ethiopia, even today ox owners are obliged to prepare the fields of oxenless landowners before their own, while those without oxen reciprocate by helping to provide feed for the livestock. In indigenous communities in Mexico, annual redistribution of plots is meant to ensure that all get an equal opportunity to farm better soil, and often here, too, families get a range of plots of differing qualities.

Even where individual fields were enclosed, as in some parts of France, communal obligations persisted, particularly in the ample additional areas where common use and shifting cultivation took place. There, farmers were obliged to open their fields to gleaners, followed by village cattle, after harvest. And they were forbidden to cut grain with a scythe to ensure that gleaners got the straw for bedding and thatch.[23] Communal governance of some sort was widespread, however the land was apportioned. The Spanish ejido, as reformulated in Mexican law after the 1917 revolution, is governed by the body of owners, who have lifetime use of a particular plot, usually passed on to the surviving spouse or eldest son. The ejido owners determine collectively, however, whether the plot will be passed down or redistributed. Irrigation arrangements may be governed by the ejido, and ejidos can join in larger collective marketing and financing efforts.

In parts of the world where richer peasants were able to acquire and expand their landholdings, while those without land resorted to sharecropping, as in colonial Vietnam or Burma, tradition prescribed that landlords could not exact rent in kind that impaired a family's ability to feed itself. The same sort of obligations pertained to clan-owned land in China, where poor relations without ownership rights nevertheless depended upon the family or clan for access to land. The breakdown of these arrangements as capitalist markets encouraged landlords to demand fixed rents from tenants often led to revolts and, eventually, revolutionary movements.[24]

These sorts of arrangements among peasant societies undoubtedly have their roots in ancient practices among hunter-gatherer peoples. Among the indigenous peoples of Northern California, careful attention was paid to the distribution of resources. In some groups families were accounted "owners" of particular oak trees or stands of oak, collecting the acorns of these trees for family subsistence. But owners had an obligation to non-owners to share the bounty as much as possible. Similarly, along the Klamath River and elsewhere on the salmon runs, tribes were careful to let a certain large number of salmon pass upstream, around the weirs they used to trap them, so their upstream neighbors, even otherwise hostile communities, could share in the chief subsistence "crop" of the region.[25]

These practices bespeak the deep-seated sense of fairness that seems to be a human attribute, and one essential for survival in community, as we will see in chapter 8. The same sense of fairness underlies the democratic character of decision making in small-scale traditional communities around the world. Democracy was not, after all, invented in ancient Greece. The Greek institutions simply transposed far older village and tribal practices onto an urban scale, not always successfully. Village and tribal democracy was probably never flawless, and autocratic societies evolved out of its breakdown. Inevitably, richer individuals and stronger personalities could dominate deliberations, just as in any rural town in the United States today. But a strong culture of reciprocity often restrained them, and often sponsored relative equality of access to land.

Finally, the web of obligations that surrounded access to land in traditional communities often include, even today, duties to care for the common properties of the community. Oaxacan immigrants to the United States periodically return to their communities in rural Mexico for communal work on roads, soccer fields, the *casa comunal* (community center) where meetings are held, or other projects in order to maintain their membership in the community and, with it, their access to a house and plot in the village and perhaps some farmland. The *tequio*, or communal work party of Latin America, like the *faena* of Spain, was part of the glue that held the community together but also a means of mutual aid. Self-help of this sort contrasts sharply with the bureaucratic and political entanglements that dependence on government public works projects generally entails.

What makes these practices relevant today? Could they provide hints or patterns for emulation to build a more resilient future? On the deepest level they are examples of the sort of culture of reciprocity and sharing that has to undergird any sustainable future farming. They also suggest methods of sharing that might work for us, even under the aegis of modern private property law. Finally, they point to the importance of sharing not just arable land but the resources of a larger land base. We'll look at the culture of resilience in chapter 8 and the possibilities for modern commons in chapter 7. In the rest of this chapter, we consider methods of achieving wider access to land that might work in our current and immediate future situations.

## Beyond the Independent Homestead

For most of us the very image of farming that we cherish involves individual ownership of an independent homestead. The peculiar English history of private property and its entanglement with notions of individual freedom and citizenship came to the English colonies as a promise of emancipation for hardworking farmers who could earn a place among the landed gentry and gain citizenship rights just by owning a piece of land. The popular image of the frontier experience deepened the positive connotations of landownership. The hardy pioneer carved a homestead out of the wilderness with toughness and ingenuity. And Thomas Jefferson made the class of homesteaders and independent small property holders the foundation of his imagined republic of virtue.

The reality of landownership in the United States was rather different, with speculators claiming much of the land opened up by federal efforts, indebtedness weighing down generations of farmers, and the biggest property holders rarely letting go their grip on power. And our literature soon added another twist to the picture of the independent settler: the desperate and lonely farm family working out a disastrous destiny in isolation from any human help—a picture with too much truth to deny for too many families over the four hundred years of American settlement.

The English model, repeated in white settler colonies throughout the British Empire, was largely a novelty in human history, as we have seen. Other models for access to farmland have dominated the history of human uses of the earth's resources. And many of them promised not just land and independence to the family farmer but immersion in a community and a continuing say in how access to land was distributed. Even in the most private-property-dominated nation in the world, the model of the independent farmstead did not rule out older forms. American farms in the past were often passed down in a family solidarity that mirrored the clan and community solidarity of traditional societies. Today this is rare. The Amish practice mentioned at the beginning of this chapter is the exception. In much of the mainstream farming community, young farmers who have family land at their disposal often face the same prohibitive problems of buying the land as those without such connections. As we saw, families often include diverse claimants to the land, many of whom have no interest in farming or in preserving the ancestral land; when one or more of these wants to sell, the member who intends to farm often has to come up with market price for the land. The children of Donald Trump may get a leg up from their wealthy father, but I know too many American farm elders and owners who are convinced their own children have to make it on their own.

Even without these attitudes, single-family ownership poses obstacles to succession. When it comes time to pass on the farm, how many farmers are willing to stay in the family house to watch offspring or newcomers transform the farm in ways they might not have wanted to see it transformed? How many are willing to pull up stakes and move? These considerations can be as compelling in questions of farm succession as the purely financial ones of sharing out the proceeds of land sales among nonfarmers with a claim on the land.

The independent homestead of our imagination affects where and how we are willing to farm, as well. We picture a home close to the barn, with crop fields and hoophouses not so many steps away. Offered a piece of land they cannot build on, many young farmers turn away, wanting "our own farm," not just a piece of land. Urban farmers have overcome that prejudice, but it remains strong among us. For example, many of us regard restrictions on building on agricultural land, such as Britain's or Oregon's land laws, as an obstacle to the restoration of widespread small-scale farming. Those laws may be too restrictive, but other cultures refused to build housing on good cropland if they could avoid it. Looked at with a wider lens, the bad habits of individualized ownership have carried our society to an extreme, gobbling up farmland for urban, suburban, and exurban development at an alarming rate.

Traditional land tenure arrangements around the world avoided our succession struggles and land loss, making redistribution of land easy by concentrating housing in nucleated settlements surrounded by their farmland. Whether owned by the community or by families, the land was easily redistributed because it did not bear the burden of the home. As we saw, land use could be strictly regulated by village custom, requiring, for instance, that even individually "owned" parcels be opened to communal grazing after harvest. In much of the world today, this system of land use has evolved into policies that restrict urban and suburban sprawl and preserve farmland much more effectively than we have managed to do in the United States. Though land concentration has occurred throughout Europe, smallholdings still predominate in many parts, and land is passed down and preserved in farming from generation to generation.

We might recover some of the best parts of that heritage by building on existing institutions and rural housing patterns. In many parts of the United States, for example, especially along back roads in rural areas, small strung-out hamlets are common, with each home occupying one to five acres, all backed by huge fields owned by others. Such settlements could be the basis for a commons of farmland available to local residents, and it could get around the restrictive zoning codes meant to protect agricultural land in states such as Oregon, where parcels eligible for construction of a house are prohibitively large for most new farmers. Brian Donahue, who spent several years farming in the commons of Weston, Massachusetts, proposes building on the old New England commons tradition, extending the pasture, forest, and wetlands still under town ownership in that region and encouraging community farming or gardening of parts of it.[26] Some contemporary co-housing projects and intentional communities already mimic the older patterns, setting aside woodlands and fields for farming and other uses.

## New Paths

Although we have few legal or organizational resources that can ensure access to land for the upcoming generation of farmers, some possibilities have begun to open up. While even older, successful farmers have to rely on rented land in many parts of the country for much of their production, this reality can provide an opportunity to beginning farmers. Farmer and writer Mike Madison recommends that aspiring farmers lease land near their markets, forgoing ownership in favor of long-term, secure leases while they build experience and capital. Madison has also pioneered a succession strategy that gives young farmers favorable leases on portions of his own land.[27] This is one among a number of approaches advocates are finding today to address the issue of access to land.

Another alternative is community solidarity of one sort or another. Small but important examples are proliferating. Urban agriculture has overcome the bias for the individually owned farmstead, finding opportunities to farm through all sorts of arrangements with private owners and public agencies. Urban farmers live where they can and commute to the farm, like traditional village-based agriculturalists, most often combining farming with other occupations. And they often share in farming with other like-minded people.

Whereas urban farmers seek permission from owners of abandoned lots, content with long-term rental agreements or concessions from city governments, farmers committed to the individual enterprise model have turned to crowdfunding in increasing numbers for help with down payments or special projects, with mostly disappointing results. Most crowdfunding efforts depend on having a sizable population of contacts to draw on, because most funding comes from friends, family, and their contacts. This is certainly community solidarity, but often it just isn't adequate, and a small farm start-up can rarely attract the sort of angel donor needed to make a land purchase.

Slow Money is like crowdfunding in drawing on a larger crowd to fund worthy projects, but it focuses on investment rather than outright donations. Slow Money describes its mission as "To catalyze the flow of capital to local food systems, connecting investors to the places where they live and promoting new principles of fiduciary responsibility that 'bring money back down to earth.'"[28]

As of 2017 the nonprofit has invested over $56.6 million in small, local food enterprises across the United States. Most of this money, to be sure, has gone to specific projects, not directly to buying land, and only a small portion has gone to farmers. But it has enabled at least some farmers to build on initial access to land with profit-making enterprises. The same is true of Kiva Zip, a crowdsourcing platform that gives small no-cost loans to farmers with as much as ten years to pay them off.

England's Ecological Land Cooperative represents a more robust approach. With major capital investments by large and small donors, the cooperative has purchased land and let it out among a growing number of farmer-members, who collectively have 50 percent of the vote in the cooperative. The Poudre Valley Community Farms cooperative, near Fort Collins, Colorado, like the Ecological Land Cooperative, has adopted classic cooperative organization with some special twists. Most members are consumers who have invested $1,000 in the co-op in exchange for access to CSA privileges on the farms supported. Dividends, distributed each year, are based on the level of spending on farm products, as in any consumer co-op. Meanwhile, a special class of investors can expect minimal returns on much higher levels of investment. Finally, Poudre Valley has agreements with state and local agencies and land trusts that will make land purchased more affordable through agricultural and conservation easements.

Our Table, in Portland, Oregon, is a cooperative that extends the co-op model to farm management. The co-op owns fifty acres on the outskirts of the city. Workers on the farm are member-owners, with 60 percent of the representation on the co-op board and a corresponding claim on dividends. Associated independent suppliers provide everything from fish to body care products for the farm store and CSA, and as members they control 20 percent of representation. And CSA members may choose to become co-op members as well, with a 20 percent share of board seats. Capital investors have no voice in the co-op as such but have an equity-based share in its assets.[29]

For the beginning farmer, incubator farm projects throughout the United States provide farmland bought with grants and donations to applicants at favorable rents, often with business training and mentoring as part of the package. In Burlington, Vermont, the Intervale Center, one of the oldest of these programs, leases land, equipment, greenhouses, irrigation, and storage facilities to small independent farms. Though most incubators require farmers to seek their own land after a period of years, others continue to make land available under relatively secure and low-cost leases. The National Incubator Farm Training Initiative profiled twelve incubators active in 2013; their website recently identified eighty-five incubator projects in thirty-eight states.[30] Even organizations with small parcels of land can accommodate one or more beginning farmers, depending on scale of operations and crops. And incubators can provide many of the resources that traditional communal arrangements also provided, as the Intervale example shows. Ensuring long-term security of land tenure, though, is a further leap. Many incubators now work with other agencies to help place and finance farmers who have graduated from the incubator.

Land trusts can help ease the transition from retiring farmers to newcomers. By buying conservation easements on the land, they can provide retiring farmers with needed income and lower the overall price of the land for newcomers. Most land trusts are more focused on environmental protection than agriculture, however, and few have the resources to make a big impact. The Sonoma County Agricultural Preservation and Open Space District, a governmental agency, was established by voters in 1990, with a quarter-cent sales tax to support land purchase and preservation. Relative to private land trusts, it has a much bigger budget and a greater impact, but, like most of these organizations, it has no program to directly offer farmland to aspiring farmers.

The most important function of conservation easements on farmland is to keep land in agriculture and lessen the impact of market-driven development on prices. Because this does not always work, some land trusts with an interest in preserving agricultural land go further. Equity Trust is a land trust that has taken a special interest in preserving farmland and has developed innovative models for doing so, from the usual conservation easement to full purchase of the land, which farmers then acquire through long-term leases. The lease agreement for the Indian Line Farm, sponsored by the Schumacher Center for a New Economics and partners, gives farmers equity in not just their buildings, but the soil itself. A soil sample was taken at the start of the lease and another will be taken if farmers decide to move away. The farmers are entitled to the equity generated by any organic improvements to the soil on the land, in addition to improvements they make to the buildings. The innovative Cuyahoga Valley Countryside Initiative is a partnership between a land trust and the National Park Service, which together renovated chosen farms within the Cuyahoga Valley National Park and have leased them to farmer-stewards, with leases of up to sixty years.

## Looking Forward

Two broad trends are likely to shape how and where we farm over the next several decades. Climate change is already causing falling yields on prime temperate-zone farmland, and not just there. Rising temperatures and carbon dioxide levels are projected to further depress yields in standard crops at a time when a growing world population is demanding more and more food. Farmers will respond by putting ever more marginal land into production.

At the same time, we can foresee rising farm costs, especially for fuel, petroleum-based fertilizers and pesticides, and phosphorus. The newest technologies are being designed to address both problems. Rising costs of petroleum and petroleum-based inputs are already pushing farmers into acquiring more and more expensive technological fixes like precision planters that inject just the right amount of fertilizer and herbicide along with each seed, doing in one pass of the tractor what has required three or more until now. The advent of electric tractors will be just another part of that trend. As in the case of technological revolutions in the past, this one will be available first and foremost to the farmers with the deepest pockets. Those who can afford such toys will continue to outcompete technological laggards at the mid- to large-scale farm size. Most farmers will not be able to keep up, or will go bankrupt trying. Further consolidation of land holding seems likely, and the middle will be further squeezed out of American agriculture. The coming years will see increased concentration of landownership, not redistribution. As long as industrial farming remains viable, alternative agriculture will continue to depend upon cheaper marginal land on the edges of farming country.

The independent homestead, patriarchal model of farm ownership that is the norm, with its roots in the peculiar history of private property and the legal constraints that favor it, will continue to stymie efforts to broaden ownership and overcome the problem of succession. Private property, of course, is our way of securing not just access to land but the sort of long-term care for the land that Wendell Berry insists is essential to developing the heritage of stewardship and "culture of affection" that underlie a conservative and regenerative farming. However, long-term "ownership" of one sort or another is only a necessary, but not sufficient, condition for using the land well. And private ownership can lead to abusive practices and the conversion of farmland to other uses.

Ideally, we will turn to traditional patterns of land tenure in order to secure small-scale farming for the future. As Brian Donahue writes, "Some kinds of land are best held privately, some commonly, and some with a mixture of private and common rights. We can think of the common interest as ranging from regulations that limit the way land may be used, through easements that convey restrictions permanently to the community for safekeeping, to outright common ownership."[31] Residential uses should be confined to narrow bands, outside of farm- and forestlands.

Farmlands may be privately owned but hedged with easements that protect the community interest in continuous agricultural uses. And a substantial portion of forested land should be brought under community stewardship. This may sound utopian, but it is all achievable under present legal constraints, provided we have the community will and the funding—no small requirements.

Until and unless we can enlarge the schemes of land trusts and community ownership that today embody this sort of vision to include even our best farmland, we will have to look to the margins, to urban lots and backroad mini-farms, to the foothills and flood-prone outlands for land to farm. But as we do so, we should be looking to models of cooperative ownership to ensure that land is broadly shared and easily acquired by yet another generation of farmers. In the process it will be important to consider the models of organization and settlement developed among traditional agriculturalists around the world, lest our efforts at communal ownership dissolve in acrimony like those of the informal communes of the 1970s, or suffer the fate of the agricultural co-ops in this country that have fallen into the hands of profit-oriented managers or have sold out to agribusiness. There are models of decision making that are inclusive and nonhierarchical, but that get real work done, which is what anyone committed to farming values perhaps most of all.[32] Key to farming for the long haul under community or cooperative ownership will be systems of governance designed to guarantee long-term, continued access to land for the farmers who need it, with the possibility of passing that land down through the generations.[33]

Underlying both traditional practices and the handful of contemporary models is a commitment to fairness that is essential to community resilience. But what most of the models used today lack are principles and practices that can ensure equitable redistribution of land, generation after generation. Traditional societies took these for granted, though they were undoubtedly developed painstakingly over time. Even the best of contemporary arrangements too often rely on the goodwill of a self-appointed (nonprofits) or elected (cooperatives) board. The Ecological Land Cooperative and Our Table at least provide for a significant farmer vote on major decisions, balanced by a community of interested nonfarmer members. Whatever the legal form, principles of distribution and redistribution need to be spelled out explicitly and in such a way as to encourage the growth and endurance of small farms of a size that are likely to succeed in the coming years. These go beyond covenants regarding use of the land like those included in many incubator farm arrangements, to include explicit commitments to fostering small, diversified farms, supporting them in hard times, and seeing to it that they are passed on to competent and like-minded farmers.

A community-owned land trust committed to providing land to small farmers might benefit from a double structure of governance. An elected board with broad representation from the local community could be balanced by a board of farmers, some with land under the trust, others independent, who would be the gatekeepers for new farmers in the trust and would have veto power over any decision to dispose of assets. Such an executive board might also serve to apportion access to commons among farmers on the community land and manage redistribution of parcels among farmers as proposals for expansion or contraction arise. Agreements might mirror ancient provisions for gleaning, common grazing, and common management of irrigation resources. Lifetime leases at favorable rents would encourage investment, while rental income from land and shared resources such as tractors, barns, and storage facilities (as at Intervale) would help cover land costs and pay for staffing.

A commitment to preserving land in diversified and productive farming also needs to be enshrined in laws that, up to now, have favored the rights of owners over the needs of community. Land reform of the sort practiced both by the United States in Taiwan and South Korea following the defeat of colonial Japan in 1945 and by revolutionary regimes since then is unlikely in this country; and it would provoke the sort of social upheaval unlikely to benefit the small farmers it might be intended to help. But land redistribution at the local level in the wake of serious financial and economic collapse might be possible. Communities might then reclaim the commons, along the lines suggested by Brian Donahue, and establish principles for access to land that could underwrite a resilient small-scale agriculture. Those principles would have to contend with three centuries of property law adverse to sustainable farm ownership, as well as the romance of the independent homestead, but the circumstances of the future might just enable a rewriting of even so fundamental a part of the ruinous market economy we face today.

Chapter 5

# Soil, Civilization, and Resilient Farmers Through the Centuries

However secure our access to land, farming will disappear from the face of the earth if our soils are abused. And abuse them we have. The evidence is overwhelming: Farming has transformed the face of the earth, and not for the good. From the inception of plow agriculture, farming has brought the degradation of soils, continuous loss of topsoil, and, at the extreme, desertification. The civilization of Mesopotamia rose with the cultivation of rich alluvial plains in a semi-arid climate. It ended when salinization from centuries of irrigation had poisoned soils, and the canals that brought life-giving waters became clogged with silt from farming-induced erosion in the hills of Armenia. Logging for masts, firewood, and agricultural expansion had eliminated the famous cedars of Lebanon by the third century BCE and stripped most of the Greek peninsula of topsoil a century later. Classical authors from Plato on deplored the effects of farming on the precious resources of the soil, but farming throughout the Mediterranean continued to deplete the stuff on which it thrived in cycles to the end of the Roman period and beyond.

Similar experiences marked the end of the Anasazi culture in the American Southwest and the decline of Mayan civilization in Middle America. Erosion in the birthplace of Chinese civilization gave the Yellow River its name and contributed to centuries of siltation. Heroic efforts to prevent flooding through dikes and dams eventually raised the river level as much as fifty feet above the original alluvial plain, and the rich silts that periodically overflowed the dikes sustained much of subsequent Chinese farming. But in the highlands of the watershed, scarcely any arable land remains.

In each case subsequent cultures eventually learned methods of soil conservation that slowed and in some cases reversed the process of degradation, but many areas of the globe were lost irrevocably to erosion, stripped to bedrock, or desertified. Learning was often uncertain, never universal, and could be reversed under changing economic conditions. We know from Roman literature that some farmers were underlining these lessons even as the destruction continued, a history similar to our own, where topsoil loss has been accelerated by industrial farming methods even as government scientists and agrarian advocates have decried the practices that are systematically destroying the bases of our own civilization.[1]

In *Dirt: The Erosion of Civilizations*, David R. Montgomery lays out the dreadful story in detail. Through much of the book, Montgomery puts the blame for the periodic stripping of the soil on which agriculture depends on population growth resulting from the success of agriculture itself. In doing so he joins a long line of archaeologists and historians who have fallen back on the notion that the population growth that agriculture enabled led to increasing pressure on the land. And as marginal land was put into agriculture, erosion followed, often with disastrous consequences for urban civilizations that depended on steady sources of foodstuffs. Montgomery quotes studies that turn to this argument for early European cultivation in Bulgaria starting fifty-three hundred years ago, in the Mayan lowlands, central Mexico, and the American Southwest.[2] In each of these cases, we have no records of just how farmers fit into the larger society, so the simple, physical explanation for the expansion and contraction of populations and accompanying soil erosion seems plausible. The trouble with the explanation, even if plausible and possible in some cases, is that traditional societies generally have no problem sustaining themselves comfortably on their land bases, and they find ways to limit their population to ensure the sustainability of their livelihoods. From cultural practices promoting sexual abstinence to herbal birth control or abortifacients to infanticide, traditional societies have known how to control reproduction. Thus, if some agriculturalists have families too big to support on the land available, we have to suppose that sensible reasons are behind the practice. These range from efforts to counter high infant mortality rates to labor shortages in the prevailing farming system.

Growing population certainly has an effect on how intensively land is farmed. Economist Ester Boserup starts with the fact that both the foraging of hunter-gatherers and the forest swidden agriculture that was probably among the first techniques for cultivation of the soil require relatively little labor. Hunter-gatherers and practitioners of swidden agriculture on forestland can provide a comfortable living on their own terms in just a few hours a day.[3] Why then turn to more labor-intensive forms of food production? Population growth certainly seems to be implicated. By the early Holocene (11,500 years ago), humans occupied all the earth's habitable regions and may well have reached the limit of the planet's carrying capacity for peoples using extensive hunting-gathering techniques.[4] Similarly, there is plenty of recent evidence that increased population has limited possibilities for the "long fallow" (twenty-five years or more) necessary to cultivating forestland; under these limits groups turn to "bush fallow," returning to the same plot long before trees mature. But this requires more intensive working, and it brings a greater threat of erosion.

Plow agriculture is more productive per acre, but far from relieving labor pressure on the population, the turn to the plow increases it. Farmers must care for their animals as well as their crops, and may be forced to cultivate more ground to feed the livestock. The land needs to be carefully prepared and the crop protected from weeds and pests. Thus, as population grows, food needs can be met with an intensification of agriculture, but this requires more labor and more mouths to feed. A growing population may provide the additional labor needed, but the evidence is that settled agricultural societies have always required more labor to feed themselves than older forms, whether hunter-gatherer or shifting agriculture. The eventual turn to plow agriculture had a lot to do with destruction of soils in many parts of the world. But the population growth that spurred this innovation does not by itself necessarily lead to abuse of the land, because, again, most human groups have known how to limit population growth in the face of scarcity. And periodic warfare among neighboring groups, so pervasive in much of the world of hunter-gatherers and tribal agricultural societies, could both limit population and provide additional territory to carry on less intensive food production.[5]

A more important factor in growing pressure on the land is likely to be the larger social context that farmers face. In this respect, three factors stand out historically: slavery, debt, and conquest. Boserup points out, for example, that one solution to the need for more labor to support a growing population is slave raids on neighboring groups, who would be forced to do the labor locals found distasteful. This was particularly true where urban civilizations grew up dependent on the surpluses agriculture could produce. Boserup puts it starkly: "It is a fair generalization to say that all the ancient communities which applied intensive systems of land utilization used servile labour, usually captives of war and descendents of such captives. . . . Where population is sparse and fertile land abundant and uncontrolled, a social hierarchy can be maintained only by direct, personal control over the members of the lower class."[6] In both eighteenth-century African kingdoms and the paddy states of Southeast Asia, we know that this was the case. And recent evidence from Mesopotamia depicts the earliest states as hungry for labor, supplied through slave raids, military conquest (with slaves as the booty), and resettlement of whole populations.[7] Intensive cultivation in these cases, as in imperial Rome, the Caribbean, and the American South, was carried on by workers with little incentive to care for the soil they worked or to develop long-term solutions to agricultural practices that drained the land. Montgomery finds this form of ownership and land use repeatedly associated with careless farming practices and consequent soil degradation, whether the owners were the late Roman aristocracy, antebellum southern planters, or corporations employing mechanized agriculture today.

Urban civilization also perfected the institution of debt to extract surplus product from a farming population otherwise intent on subsistence. In ancient Sumer, the first agricultural civilization, both temple officials and merchants kept elaborate records of debts, the first focused on temple tithes, the second on payment for goods delivered. Agriculturalists fell into debt first of all, it appears, under the compulsion of tithes owed the great temples of Sumer's cities. But specialized production of household necessities also enabled the growth of a merchant class to whom unwary farmers could become indebted. Whatever the source of the debt, it could lead to increased pressure on the soil base, because the consequences of nonpayment included the loss of servants, children, wives, even personal freedom—and, of course, land to the holder of the debt.[8] In this way, undoubtedly, pressure on the land increased. And in this way, too, we know that land passed into the hands of absentee landlords and was worked by slaves, with the consequences mentioned already.

Modern colonial practices demonstrate how conquest could distort a functioning and ecologically appropriate agricultural economy, as Montgomery also shows. In the Sahel a centuries-old reciprocal relationship between nomadic herders and farmers practicing a conservative shifting agriculture was disrupted by French colonialism. The French, anxious to create a money economy and employ locals in producing goods for export, imposed hut and animal taxes, which required both nomads and farmers to seek sources of money income. Confined to smaller territories by new administrative boundaries, herders and farmers intensified production on the land remaining to them. Subsistence farmers, meanwhile, were being displaced by French planters growing cotton and peanuts for the European market. Degradation of the soil followed closely behind these new social and economic relations, exacerbated by rapid population growth with the introduction of modern sanitary practices. Starting in 1972 successive droughts wreaked havoc on a now precarious eco- and social system, where drought had always been gracefully weathered in centuries past.[9]

Complex causes lie behind the population boom and agricultural expansion into marginal land at the onset of the Qing dynasty in China (1636–1912). The banning of female infanticide, a network of granaries for times of shortage, the world's first smallpox inoculation program, and a tax freeze on cropland all contributed to better health and a population explosion from perhaps 150 million to 300 million people. At the outset of the regime, moreover, the eastern seaboard was cleared of population as a buffer against foreign influence. As this population moved westward, it displaced smaller ethnic groups, who began farming rented land in the hills and mountains. The Qing also encouraged ethnic Han to migrate into mountainous Sichuan and Shaanxi Provinces. These areas, not suited to the staples of the Chinese diet of rice and wheat, would not have sustained the new populations if American crops—maize, potatoes, and sweet potatoes—were not available for adoption.[10] But those crops, particularly maize, probably contributed to subsequent erosion of the new hillside farms.

Central to these stories are social structures that give some people power over others. A striking feature of many traditional societies is their ingenuity in preventing just such a situation.

Europeans never really grasped that the "chiefs" of tribes throughout the Americas did not have the power to command, much less to sign away large chunks of tribal territory. They were spokespeople because they were honored for their rhetorical abilities. To maintain a place of honor in their communities, they knew not to demand obedience. They were also expected to practice generosity and were likely to be impoverished by the demands of their people on their households.[11] In highland Burma and elsewhere among the hill peoples of Southeast Asia, headmen who became too demanding were simply assassinated.[12] And in many traditional cultures, the wealthy were periodically required to redistribute their riches, preventing an accumulation large enough to overwhelm lesser individuals and fund a real kingship.[13]

Thus it appears that unique historical twists lie behind the development of the sorts of societies that pushed farmers to exploit their soil beyond its capacity to deliver. In the case of the French conquest of the Sahel, the story is clear enough. And we see similar stories in successive conquests of Mesopotamia by societies not always equipped to appreciate the delicate balance between exploitation of agricultural riches and maintenance of the irrigation and soil resources on which those riches depended. But we know little about how and why the egalitarian tribal societies of the Neolithic world drifted into hierarchical states beyond these brute examples of conquest from without.

Traditional, subsistence-first farmers, as we have seen, are in it for the long haul. They want first and foremost to support their families through farming. Provided they have secure access to land, they do not have a short time horizon: they intend for their families to persist "from generation unto generation." And the degradation of the land base on which they depend is not an option within that time horizon. But the efforts of urban elements, conquerors, and overlords to benefit from the surpluses that agriculture is capable of producing can shorten a farmer's perspective as desperation to make a living or meet the demands of an official or a creditor overrides good judgment about how and what to farm. In extreme but not rare cases, farming has fallen into the hands of a ruling class employing slaves with little dedication to good farming practices. In a market economy the forces that demand production at all costs often shape the farmer's own needs and wants. And these may be enough to encourage practices that exhaust the land.

Technological innovations like the plow or the tractor no doubt factor into this picture, as does ignorance of appropriate farming techniques for the terrain. Though many cultures learned from the errors of the past, those errors were made by farmers sometimes ignorant of the outcome of their practices. Farmers forced onto frontier and marginal lands are particularly apt to make mistakes in management of the land, the more so if they do not have secure tenure. And since such lands are often the hillsides disfavored by the first farmers or wealthy landowners, such mistakes can have consequences for both the hillsides and the lowlands beyond.

In El Salvador, for example, peasants on coastal cattle ranches were forced out in the 1950s as landowners adopted the new pesticides that made it possible to raise crops for world markets. The peasants resettled on the hilly land that was still available to them. But they brought with them from the coastal plain practices of plowing and agricultural burning that proved disastrous on the hillsides. Concerted retraining was needed to turn around the soil erosion that was already reducing livelihoods of peasant farmers by the onset of the civil war in 1979. It may not be too far-fetched to argue that the apparently careless use of the land by European pioneers on American soil had as much to do with ignorance of appropriate techniques as with a disregard for principles of stewardship. My great-grandmother's generation, after all, was lured to lay bare the arid plains by the experts' promises that "rain follows the plow."

The received wisdom in government, academic, and agribusiness circles up until recently has encouraged careless soil management. As we will see, that is beginning to change. But the economic imperatives that accompany bad advice have not changed, and will not, until subsidized commodity production comes to an end. In the meantime, we have to find ways to subsist, and if possible thrive, while honoring our vocation to stewardship.

Farming for the long haul depends crucially on soil conservation, renewal, and restoration. It must be regenerative, in short. That is a key reason why industrial agriculture has failed us so badly. It is also a reason why even so traditional a method as tilling the soil is problematic. Under certain, very exacting circumstances, tillage may be compatible with the long-term health of the soil, though under any circumstances it temporarily disrupts what biologist Elaine Ingham calls the "soil food web" that contributes so much to fertility. But in most climates, under most conditions, tillage has contributed to the degradation of the very basis for agriculture. With topsoil losses continuing to mount year after year, an official of the UN's Food and Agriculture Organization recently predicted that, at current rates of soil loss, the world has just sixty more harvests—sixty years more, that is, of continued agricultural production.[14]

As we have seen, there is plenty of evidence that traditional farming contributed to the degradation of soils around the world and to widespread desertification by as early as the third century BCE. But there is also a rich track record of traditional soil conservation, renewal, and management practices that we can learn from. Only a few of these are common knowledge even among practitioners of sustainable alternative agriculture today. So it's worth reviewing some of the practices of the past and present that have allowed traditional agriculturalists to maintain a presence on the land for hundreds and even thousands of years.

## Conservation and Regeneration in Traditional Agriculture

Roman authors were sufficiently well-versed in the dangers of cultivation to develop a series of recommendations for soil renewal and conservation. Crop rotations, cover cropping, fallowing, and manuring were among their recommendations. Where Hesiod, the earliest Greek writer on agriculture, recommended plowing in straight lines regardless of the lay of the land, by late classical times farmers were plowing on contour and building terraces to hold back soil erosion. As cropland became scarce in Europe by the late Middle Ages and Renaissance, the Roman recommendations were revived among aristocratic landowners, contributing to the enduring landscapes of much of Europe into the modern period. And European peasant farmers were already practicing the mixed livestock and crop farming they would eventually bring to North America, with careful use of manure and fallowing and, in many parts of Europe, successful control of erosion over many centuries.

The sustainable farming practices of Native American peoples of the Midwest may stem from similar learning after the collapse of the Mound Builder culture of the Ohio and Mississippi River watersheds in the sixteenth century. Pueblo cultures in New Mexico and Arizona also may reflect learning that followed the dispersal of Ancient Pueblan civilization. In fact, the small-scale societies, sustainable agriculture, and ecological wisdom found throughout what is now the United States might be explained as learned responses to the failures of earlier civilizations. Similarly, the careful soil conservation and renewal practices of central Mexico that we will look at below followed serious erosion in the heartland of ancient Mexican civilization. And agriculture has persisted in the Near East, despite the widespread degradation of ancient and classical times, thanks to the adoption of wise soil conservation techniques dating back more than two millennia.

### TERRACING

Perhaps the most common method of soil conservation is terracing. Permaculture introduced many of us to the use of swales to conserve moisture and nurture plantings of trees. But swales are just one among many forms of terracing, and, as practiced in the permaculture world, they are primarily conceived as structures for retaining soil moisture. Throughout the world, terraces have been carefully adapted to the situation and the cropping requirements of farmers as devices for retaining moisture but also providing a surface for planting, sometimes intensively. Terracing takes many forms, from shallow depressions for fruit trees on gentle slopes to elaborate, stone-faced systems with aqueducts and rock-lined irrigation canals. Gary Paul Nabhan lists several different sorts of terracing systems, based on observations in Latin America, Central Asia, and the Middle East, including long and short bench terraces; hillside drainage canals for channeling excess runoff; terraces behind check dams and gabions; orchard terraces; balcony or window box terraces for orchard species on steep slopes; and bowl terraces for single perennial plants on shallower slopes.[15]

The width of terraces depends upon slope and intended crop and cropping system; some are wide enough to be worked with a tractor or team of oxen, others so narrow as to accommodate only one row of vines. Sloping terraces, common in Meso-America, are relatively low-cost affairs, requiring a lip, with a shallow drainage ditch below, both constructed on contour. The lip is typically planted with maguey, mesquite, or small fruit trees, and the face of the embankment with sod. Traces of such plantings have also been found throughout the American Southwest. Over time the slope of the land behind the embankments becomes more level, as plowing and rainfall spread soil downhill. The tea plantations of China often use heavy mulch rather than actual terracing, but the effect is the same.

In Mexico and Guatemala flat or bench terraces on steeper land are built by hand, digging out soil at the back to fill the forward space.[16] In some cases the face of the terrace is regularly cut away, gradually incorporating less fertile subsoil into the terrace while widening the cropping space. The risers on these terraces are usually as vertical as possible, with a lip to hold back runoff. Here, too, the lip of the riser may be planted to help retain soil. The terrace surface itself may be gently sloped back to a small channel at the foot of the riser to further reduce runoff. Or, where irrigation is part of the system, the channel at the back of the bed is wider and deeper, and the terrace itself flat. One variation of terracing on very gentle slopes is the waffle garden, a small catchment area where runoff or rainfall may gather before spilling over into the next garden below. In parts of China similar arrangements can span fields from less than an acre to many acres over large expanses of cropland, with fields devoted to paddy rice during the rainy season and other crops after harvest.

In the elaborate ancient systems still found throughout the central Andes, steep mountainsides have been narrowly terraced with stone-faced risers and irrigation channels running at their feet. These terraces provide the solar exposure the narrow valleys of the region typically do not get. The soil in them remains richer than anything on the mountainsides around them, attesting to years of incorporating organic matter into these terraces. Very old stone-faced terraces on steep land can be found in the vineyard and orchard regions of Southern Europe, in the Middle East from Lebanon to Yemen, and elsewhere.

Sloping and temporary terracing systems may contribute to erosion if not carefully maintained. And terracing on steeper slopes always risks the dangers of a breakdown under heavy precipitation and runoff from higher up. But careful design of these systems is largely successful in preventing disaster. In the case of the stone-faced terraces of the Andes, the engineering has preserved both the hillsides and the terraces over some five centuries of abandonment.

### CREATING FERTILITY

Western writers have long scorned the ancient practice of swidden agriculture, still in use today in many parts of the world. But cutting and burning forest or savanna is a means of enhancing the fertility of the soil to be worked, and where proper rotations are used the practice can be maintained indefinitely, at times even building fertility. More accurately called shifting cultivation or even land rotation, swidden has sometimes been seen as a pioneering form of farming, practiced as peoples moved into new areas. But most studies of the practice today find it an integral part of systems of crop rotation, with more intensive agriculture in permanent sites often mixed with shifting cultivation on outlying land; apparently this has been the case through much of the world, reaching back centuries. Where sufficient land was available, Chinese and Southeast Asian farmers maintained both permanent wet paddy rice cultivation and swiddening as part of a diversified farming strategy. Most West African farmers employ what Paul Richards calls a "rich tool kit" of cultivation techniques that commonly includes both shifting agriculture and permanent plantings employing manures and compost on wetter lowlands.

In West Africa farmers tailor their techniques to savanna or forestlands. In forests, land may be cleared then cultivated for one to four years, followed by a fallow of five to fifteen years. This "short fallow" system takes account of the rapid regrowth of forest in tropical climates and the abundant nutrients that burned plant matter provides. In savanna lands the period of cultivation may be much longer, up to ten years, but the fallow also is much longer, as grasslands take longer to reestablish themselves.

In both cases good farmers take particular care to protect the soil, especially in the first year. In the forests larger trees will be left standing, as will stumps of smaller trees and bushes, preserving a root structure that helps hold the soil. In some cases, as in older practices in Ghana, burning is omitted and the debris from cutting is left to rot for a year, providing fertility and soil protection for the new field. Where active erosion is a possibility, farmers employ mounds built on contour or brush weirs to hold back the water and collect sediment, and they practice minimal to no tillage in preparing the soil. Intercropping and rapid crop rotation both ensure that the soil is covered as much of the time as possible and that fertility needs are met, with less demanding crops succeeding more demanding ones as the end of the cultivation period approaches.

And farmers routinely plant pigeon peas and leguminous bushes at the beginning of a fallow period to speed up the fallow.[17]

As this look at actual swidden practices illustrates, sustainable agriculture doesn't just conserve the soil. It builds soils. Perhaps the most striking example is the *terra preta* of the Amazon. Amazonia's soil is fragile and degraded, owing to millennia of tropical rains dissolving and leaching away minerals, while the tropical heat quickly oxidizes organic matter. The rainforests make their own fertility through constant shedding of biomass. But once stripped of forest cover, the land quickly loses its fertility and even its ability to regenerate forest. Amazonian peoples dealt with these conditions in two ways. First, they avoided wholesale clearing, instead favoring tree crops that could be interplanted in the forest or on its edges. The peach palm, for instance, was domesticated through intensive breeding thousands of years ago and became a standard crop as far away as the Caribbean and Central America at least two thousand years ago. The contemporary Amazonian forest is so rich in edible plants, anthropologists maintain, because people planted them. The rainforest is actually old orchards.[18]

The second response to the fragile character of the soils was to build highly fertile and stable soils—*terra preta do Indio* as Brazilians refer to it today—thousands of square miles of dark, fertile soil, as much as six feet deep. Westerners discovered—or rediscovered—biochar from research on these soils, which are prized by Brazilian farmers and gardeners today. They are composed of carbon in the form of charcoal, impregnated with organic matter and microbes. On average there is much more plant-available phosphorus, calcium, sulfur, and nitrogen in these soils than in the rainforest. In addition they are rich in organic matter, retain moisture and nutrients far better than other soils, and host a great deal more microbial life. Though the original deposits may have been accidental results of human habitation, a culture of terra preta appears to have been spread with the migration of Arawak-speaking groups throughout the region some two thousand years ago. In some places vast settlements seem to have produced hundreds of acres of new soil. And the process goes on today. One anthropologist who spent years among the Kayapó describes a landscape of constant, smoldering fires of weeds, garbage, crop waste, even termite mounds, suggesting a conscious effort to create terra preta.[19]

The practice of making biochar and incorporating it into compost piles and the soil appears to have been widespread in North America as well, and not just among indigenous peoples. Hohokam farmers in Arizona evidently made and used biochar, even importing it to terraced lands at some distance from their villages.[20] Early European scientific studies suggested that charcoal could be a useful source of fertility, and farm trials were conducted in both England and North America in the eighteenth and early nineteenth centuries. Most found some improvement in soil fertility, and despite the complaint that the process of making charcoal was more expensive than it was worth, there apparently always have been farmers who made and used biochar. Europeans investigated biochar because the practice was ancient. Charcoal was used as a soil amendment among the Romans and perhaps the Greeks and Egyptians before them. West African farmers will use bonfire sites as nurseries for new plantings. And evidence for wide production of biochar has been found in Neolithic sites in Norway, as well.

### MANURE AND "THE SACRED DUTY OF AGRICULTURE"

The European integrated farming systems that were the basis for most of the written farming advice from the sixteenth century onward, however, relied much more heavily on manure. It seemed clear to English and other writers that farming systems that failed to incorporate enough animals would end by impoverishing the soil. European and American critics of the tobacco culture of the colonial and postcolonial mid-Atlantic and southern United States insisted over and over that the rapid depletion of the land under tobacco (and later cotton) could be remedied with regular applications of manure. And northern farmers, like their European progenitors, tended to prize a farm that produced sufficient sources of manure to keep the cropland in constant fertility.[21] This tradition persists in contemporary biodynamic conceptions of the ideal farm and in some organic practice. And critics of the new chemical-based fertilizers being developed in the nineteenth century were quick to point out the dangers of failing to replenish organic matter as a result of adoption of water-soluble nutrients. The slogan of organic agriculture that we should "fertilize the soil, not the plants" harks back to these arguments.

Manure has frequently been spread directly on soils, usually as part of fall soil preparation. But manure from animal stalls is often stacked and allowed to compost before spreading. In most of Europe and the United States, composting was haphazard until Sir Albert Howard introduced the Indore method developed for British colonial plantations on the basis of Hunza practice in the Himalayas. But the use of "hot beds" in French intensive gardening, starting at least in the fifteenth century (and probably going back much further), yielded a comparable product. The practice, described in the next chapter, used manure and animal bedding to heat seedbeds and, eventually, to force crops in cold weather. When exhausted, the beds provided a rich compost for use as a top-dressing in seedbeds and as a soil amendment.

Compost of one sort or another is used widely around the world. In Mexico and China densely settled villages produced copious amounts of waste that was stacked in layered piles of bedding straw, animal manure, kitchen waste, yard sweepings, feathers from poultry, humanure, and ashes from the hearth. In the intensively farmed market gardens of Quetzaltenango, Guatemala Mayan farmers collect oak and pine litter from nearby forests, spread it in stables for a week or more to absorb manure and urine, then put it directly into the soil. Maize stalks and other crop waste serve the same purpose elsewhere.[22]

Much of the waste shoveled into fields and compost piles in traditional societies was humanure. In China, Japan, and Korea, just before the First World War, F. H. King found a flourishing business in the sale of chamber pots full of "night soil" from urban residences to farmers. Networks of canals served as much to transport the valuable nutrient to farmers as to bring crops from farmers to the city. Similarly, in sixteenth-century London, farmers along the River Thames employed barges to transport market produce to town and bring back copious supplies of horse and human manure for their fields.[23] And eighteenth-century Parisian market gardeners valued "Parisian mud" from the city's sewage system as an important source of nutrients for some crops.[24] In China slurry from cesspools was applied directly to the soil and worked in, or it was composted along with other ingredients.

Though we know that human waste can carry disease, we also know that most of the pathogens from the human gut degrade quickly when exposed to air and sunlight or mixed into good soil. Our taboo on the subject of using human waste in agriculture means that we actually have little research on just how dangerous it is or isn't and almost none on how best to use it. The considered judgment of the British health officer for the city of Shanghai, quoted by King, is worth pondering:

> Regarding the bearing on the sanitation of Shanghai of the relationship between Eastern and Western hygiene, it may be said, that if prolonged national life is indicative of sound sanitation, the Chinese are a race worthy of study by all who concern themselves with Public Health. . . . Chinese hygiene, when compared with medieval English, appears to advantage. The main problem of sanitation is to cleanse the dwelling day by day, and if this can be done at a profit so much the better. While the ultra-civilized Western elaborates destructors for burning garbage at a financial loss and turns sewage into the sea, the Chinaman

> uses both for manure. He wastes nothing while the sacred duty of agriculture is uppermost in his mind.[25]

**FERTILE WATERS**

Cultures with insufficient or nonexistent sources of manure worked out other ways to enhance soil fertility. Adding soil from ant and termite mounds, ash and charcoal from household fires, and the widespread practice of burning the stubble left from harvest were all ways of increasing fertility in cropped soils. But some of both the most primitive and the most sophisticated approaches had to do with taking advantage of the fertility floating in floodwaters. By an accident of geography, Egyptian farmers benefited from the twin sources of fertility that Nile River flooding brought. Each year the rising waters of the Blue Nile brought mineral-rich silt from Ethiopia, while the White Nile flow was rich in organic matter from tropical forests. Farmers learned to channel the floods, spreading the waters into successive basins over large expanses of delta land.[26] Egyptian farmers supplied the cities of Egypt for thousands of years on the ever-renewing resources of their soils. The construction of the Aswan High Dam in the 1970s in a fit of modernizing amnesia put an end to agricultural abundance in the oldest continuous civilization in the world.

In central Mexico farmers likewise channel silt-rich soils from eroding hillsides, opening dikes to sheet-flood fields where the water soaks in slowly, adding nutrients to the surface. Flows are blocked with check dams of stone or concrete or even brush, depending on the depth and strength of the flow. Water may be channeled along terraces protected by a raised and vegetated lip, or it may be sent by canals to flood fields surrounded by low walls. If the field is already planted, it will be opened only long enough to irrigate and deposit a little silt. Otherwise floodwaters will be allowed to form a small lake, depositing silt as the water recedes or is drained away to another field. Farmers must often level their fields with ox-drawn scrapers afterward, as the deposits are not necessarily even.[27] The more elaborate of these arrangements are carefully governed by committees of farmers, evidently modeled on the ancient irrigation associations devised to handle river and canal irrigation in Andalusia, Spain.

In the American Southwest similar methods of managing floodwaters for soil fertility appear to date back centuries. The remains of check dams in the Chiricahua Mountains in southern Arizona and New Mexico speak of a culture that farmed this arid region through careful management of water and the deposits it brought with it. As some of these structures have been restored, researchers have found that vegetation returns over a larger area of land, springs revive, and moisture is retained in the soil much longer. More adapted to the washes and flats of arid lowlands are the uses of flooding by O'odham farmers in the Sonoran Desert of southern Arizona and northern Mexico. Brush weirs and check dams slow the water and allow it to drop silt and an assortment of organic scraps while sinking more slowly into the soil. Then farmers collect the detritus, called *abono del rio* (river fertilizer) in Mexico, and incorporate it into their fields. One batch analyzed by Gary Paul Nabhan was predominantly leaves of leguminous trees and shrubs, twigs and charred wood, seeds and decaying fruits, and droppings from several species of desert animals.[28]

Some of the most sophisticated versions of these practices are found in East Asia, where farmers have dealt with flooding and the threat of erosion with levees, canals, terraces, and narrow earthen walls around fields to trap runoff and force it to settle. The silt-laden waters of the Yellow River, carrying the remains of the soils of the ancient heartland of China, have been induced to spread and create thousands of acres of new land while fertilizing ancient fields. In China, Korea, and Japan, farmers have carefully leveled fields to allow water to spread evenly. In some areas field reservoirs were constructed to take extra runoff and sometimes serve also as fishponds. After the rains, the water is returned to the fields as irrigation, while the silt at the bottom is either spread over the field directly or added to nearby compost pits along with clover in flower, crop residue, and manure. Similar pits lined many of the canals that crossed China's alluvial plains, where silts dredged from canals could add to the fertility of the compost.[29]

The paddy rice of East Asia, Southeast Asia, Indonesia, and elsewhere, often combining complex irrigation systems and permanent terraces, exploits another source of fertility in the effects of submersion on the soil. Submerged soils quickly acquire a relatively neutral pH, regardless of whether they started out as more acidic or more alkaline. Organic matter in the paddies breaks down more slowly than in aerobic conditions, but produces more stable carbon sequestration and more available nitrogen. Other minerals of all kinds become more available to plants. Adding straw to the paddies, as many Asian farmer do, increases fertility.[30] The compost pits mentioned above must have benefited from these same effects, immersing organic matter in runoff and a slurry of canal silt to produce the same anaerobic conditions found in the paddies. Where farmers incorporate fish such as carp into the paddy culture—as they have from time immemorial in the Philippines, Indonesia, and Japan—fertility is enhanced and pests and diseases are controlled. And the paddy provides an additional valuable source of protein.

## The Future of Soil

As I noted in the last chapter, two broad trends are likely to shape how and where we farm over the next several decades: growing demand for food in an environment where overall productivity is falling due to climate change and soil degradation; and the rising costs of farming at the end of the age of petroleum. We can expect to see more and more marginal land put into production alongside increasing concentration in landholding. In the near future, small farmers more than ever will be forced to occupy marginal lands—arid, flood-prone, steep, or with poor soil quality. And where they have the advantage of better soils, they will be under heavy pressures to produce at all costs.

In this context the rich repertoire that traditional farmers have developed to manage, restore, and enrich their soils will be increasingly important. Terracing is a lost art in the United States, practiced mostly in the wine grape industry. The need to farm more intensively on marginal lands could see the revival of ancient practices. That effort will require both careful study and experimentation to avoid structural failures and the erosion that badly done or ill-maintained terracing can provoke. Gary Paul Nabhan has probably gone the furthest toward applying ancient techniques in North America. He gives detailed instructions from his own experience and the work of Ted Sheng on building bench terraces and waffle gardens in *Growing Food in a Hotter, Drier Land*.[31] Elsewhere a promising sign of change is the restoration of the ancient Inca terraces in the Ayacucho region of Peru, where peasants are learning to rebuild and put to use the waterworks and terraces that have gone unused for five centuries.[32] And Sheng himself developed a simple formula for building terraces from his own experience in Asia and Latin America. Work in riparian restoration and regenerative agriculture, particularly in the arid West, has begun to inform efforts to recapture the fertility and water-retaining capacities of the land through old techniques like check dams, brush weirs, and green barriers. And permaculture's emphasis on careful management of moisture also has lots of lessons for erosion control and soil building.

One key to soil conservation and enrichment that is gaining ground today is minimal tillage or no-till farming. Not all traditional farmers adopted the plow. Today's seed drill might be thought of as the modern equivalent of the digging stick that European settlers and colonial officials regarded as the epitome of savage ignorance. In fact, a version of the digging stick is still used in West Africa on soils particularly susceptible to erosion, and a No Till Center in Ghana is encouraging its revival. No-till extends this wisdom to all cropping, and like traditional swidden agriculturalists, no-till farmers plant in the remains of the previous vegetative cover. At Singing Frogs Farm in Sebastopol, California, Paul and Elizabeth Kaiser manage highly profitable vegetable production by using compost wisely, leaving the roots of harvested crops in the ground, disturbing the soil as little as possible, and keeping a diversity of living plants in the ground as often as possible. The Kaisers produce year-round and gross over $100,000 an acre on their three-acre mini-farm, underlining the point that farming for the long haul can be profitable today.[33] No-till vegetable farming depends on copious amounts of compost applied to permanent beds where the roots of the previous crop have been left to hold the soil and soil microbes as they decay. And perhaps we are too fastidious, insisting on a clear bed for our choice of plants, whereas other agricultural systems tolerated and even encouraged certain weeds for the benefits they brought cultivars. Bob Cannard's Green String Farm is famous (or notorious) for tolerating weeds among the vegetables and cover crops, nurturing those that add fertility or enhance soil quality while cutting down those that crowd or compete unduly with cultivars.[34]

Current trends in pasture management likewise mimic versions of herding that were closer to natural conditions. As we saw in chapter 3, holistic management, pioneered by Allan Savory, relies on bunched herds of livestock briefly occupying a paddock where they have the opportunity to graze pasture plants to optimal levels, then move on. The controlled grazing, the animal manure, and even the churning of the soil by livestock hooves stimulates the growth of the most desirable plants. Joel Salatin has developed similar techniques for the more temperate climate of the eastern United States. The results for soil health have been so impressive that a "carbon farming" movement has become the focus of the new regenerative agriculture.

As the new soil science shows, healthy soils are based upon a healthy microbiology interacting symbiotically with plants. One result is sequestering carbon in the soil, with perennial grasses particularly adept at nurturing mycorrhizal (fungal) networks that store carbon along mycelial threads spread throughout the living soil. The deeper roots go, along with their fungal companions, the deeper the carbon is stored, making deep-rooted perennials like prairie grasses and trees particularly good at sequestering carbon. Some claim that if 25 percent of arable land were farmed or ranched using no-till methods, all the excess carbon currently in the atmosphere could be returned to the soil. Though annual crop production is less effective at carbon sequestration than perennials, no-till methods in annual cropping can contribute to countering excess carbon buildup in the atmosphere as well.[35]

No-till practices in grain production have until recently depended upon herbicides to hold down weeds. But researchers and farmers in the upper Plains states have reintroduced cover cropping and complex rotations to restore soil health and manage weeds and pests. David Montgomery reports that three-fourths of South Dakota farmers have embraced no-till, just since 1990. Many of them practice what Montgomery and others call "conservation agriculture" involving: "(1) minimum disturbance of the soil; (2) growing cover crops and retaining crop residue so that the soil is *always* covered; and (3) use of diverse crop rotations."[36] North Dakota rancher Gabe Brown has turned to multispecies cover cropping and added intensive grazing to the mix. Brown has become a poster child for regenerative agriculture on a large scale. For Montgomery, chronicler of agriculture's long-term devastation of the world's soils, the new movement is reason for hope that this history will not repeat itself. The evidence makes clear, as he puts it, that "it's possible not only to restore soil on a global scale, but to do so remarkably fast."[37]

Perhaps most promising of all, the long-term goal of plant scientist Wes Jackson and The Land Institute to develop perennial grain crops is being realized. Kernza, a perennial cousin of domesticated wheat, is now being commercialized ten years ahead of schedule. It doesn't require replanting and hence eliminates the soil loss associated with tillage, and like all perennial grasses, it builds soil and sequesters carbon. The Land Institute is working on perennial oilseed crops and legumes as well, with the ultimate goal to mimic the prairie's perennial polycultures while producing abundant food for humans.[38] The perennial crops that are the goal of The Land Institute's breeding program, like the tree crops at the core of Mark Shepard's restoration agriculture, promise both the virtual end of soil erosion and the long-term soil building that geographer J. Russell Smith long ago envisioned as a "permanent agriculture."[39]

### FERTILITY: DILEMMAS AND SOLUTIONS

For crop farmers fertility has to come from somewhere, and many of us import it. Off-farm sources of fertility have lately gotten a bad rap from some advocates of biodynamic and organic farming. But the truth is that any farm that ships product off the farm has to be able to find ways to restore fertility. No-till methods are one way. But even some of the ranches that have pioneered the carbon farming movement are bringing compost and other nutrients to their pastures.

Traditional societies often went to great lengths to gather and process urban waste, transporting kitchen and food-processing waste and manure long distances to bring fertility back to the farm. We've only begun that process, with municipal composting beginning to spread in the United States and the processing of sewage sludge for use as fertilizer advancing. We are hampered by the contamination that municipal compost and sewage systems have to accept. Even garden and yard waste from non-organic sources are potentially contaminated with long-lasting herbicides that can be fatal to crops, despite composting. And sewage sludge, even stripped of heavy metals, potentially includes pharmaceuticals, body care products, and other chemicals that have unknown effects in the soil. At the same time, there are growing concerns about further mining of mineral resources for farming, particularly the apparently limited supply of rock phosphate. Similar concerns have been raised about overharvesting seaweed, used as a source of trace minerals by many market gardeners.

The result is that we have to be both selective and eclectic about off-farm sources of fertility and better exploit readily available sources. A good starting point might be to re-read Eliot Coleman's discussion of sources of fertility for the farm.[40] There are common sources of fertility most of us neglect. Will Bonsall, for example, recommends a judicious application of wood ash on grasslands and pastures; it is high in calcium and potassium but also quite alkaline. He uses hardwood chips as a source of fungus-based fertility around fruit trees, berries, and other woody perennials.[41] And many organic farmers have long advocated using leaves from our own woods or municipal collections as a significant source of nutrients. Others use commercial food wastes, from coffee grounds to scraps from restaurants, school and hospital cafeterias, grocery stores, and food processors as feedstock for compost.

Most of us probably under-exploit the potential of cover crops and green manures as well. Though a standard part of the repertoire of organic farming, cover crops often have taken a backseat to concerns for maximum productivity and minimum effort in preparing beds. It's easier to apply generous layers of compost—no matter how laboriously (or expensively) acquired—than to dig in a cover crop. At least that is the message of a lot of recent market gardening books. But cover cropping plays an important role in some vegetable growers' systems, and it is an essential part of rebuilding fertility for grain farmers like Gabe Brown, whose work regenerating his North Dakota ranch was mentioned above.[42] Cover crops can stack functions by setting nitrogen in the soil, assembling plant-available nutrients, protecting bare soil, providing forage for animals or a crop for human consumption, and serving as a seed crop. The balance of such usages will depend on our aims, since cropping reduces the nutrients available to the soil, but the versatility of cover crops has recommended them to farmers through the centuries. While many farmers have traditionally sown cover crops among grains or between main crops, Eliot Coleman has extended the practice and undersows green manure crops in the vegetable garden. He gives specific recommendations for timing and varieties in *The New Organic Grower*.[43]

In the end there will be no sustainable agriculture without humanure. The highly processed and still-dubious products of municipal sewage plants may not suit vegetable production, but they could be well used, and are being used, despite National Organic Program objections, on grain and pastureland, and in orchards as well. We can look forward to the day, coming quickly, when the microbiome theory of health will become the dominant paradigm and put the germ theory of disease in its place.[44] And perhaps one day public health authorities and the "food safety" industry will take account of significant actual risks, and not merely possible ones, in formulating rules and regulations for food producers. Small-scale fruit and vegetable producers may have access to less contaminated product than municipal plants, but it will take a public health revolution and a lot of research to make humanure a part of the vegetable producer's repertoire, despite the experience of centuries of farming.

The new soil science has helped us understand the importance of microbiology in fertility, confirming the insights of Sir Albert Howard seventy-five years ago. If we now know a bit about the ways in which microbes engage in complex biochemical exchanges with plants, mining soil nutrients and sequestering carbon from the atmosphere, that knowledge has served to bolster two lessons for practice. On the one hand, it underlines the importance of disturbing the soil as little possible, contrary to centuries of the practice of tillage. On the other hand, it has confirmed what good farmers have almost always known: that it is supremely important to build soil organic matter in whatever way possible. Traditional practices, from controlled flooding to manuring to cover cropping, have been validated by the new science—not that good farmers ever needed such validation. And we can now appreciate better the use of biochar, composting, and paddy production in continuous soil fertility. Whether we will really need the rash of new commercial products advertised to enhance the microbiology of our soils is more doubtful than our need to heed the lessons, good as well as bad, of traditional farming.

Resourceful farmers, it turns out, learned a great deal that modern agriculture has forgotten, even if the process of learning sometimes sprang from disastrous experiences of soil loss. In the next chapter we look at more of the solutions that resourceful farmers have devised to cope with climate challenges and to manage water resources.

CHAPTER 6

# Resourceful Farmers

Farmers are jacks of all trades, masters of multiple tools. They are also intimately tied to the resources that enable them to farm, from climate to soil to water to energy. The best farmers have learned to manage those resources for sustainability, and they tend to prefer the resources at their immediate disposal to those imported from elsewhere. Our repertoire of organic and permaculture techniques depends heavily on what we have learned, even as we adapt our farming practices to what works for us in our particular circumstances. But what we have learned is a limited set of techniques developed in a few places and passed on by a handful or two of pioneering mentors and writers. Hungry for help, we are attentive to the latest technique, scientific finding, or tool that can help us farm better. But we have left behind much that farmers through the centuries learned and employed in their own struggles to bring food to the table, and too often the latest and greatest just adds to the expense of farming.

Most of us operate with only a small toolkit for managing resources. Organic farming has enlarged that toolkit, and permaculture has added approaches derived from traditional farming techniques. We know how to make compost (thanks to the Hunza farmers who inspired Sir Albert Howard), often rely on permanent beds (like a great many indigenous farmers), and use mulch to conserve water and build soil tilth. Permaculture has popularized swales built on keylines (contour) and some wise water-conserving design tools, as well as multilayered and diverse forest gardens based on indigenous and traditional models. Traditional agricultural systems offer an even richer diversity of adaptations that we will need to draw from to remain resilient in an uncertain world. In this chapter we will look at adaptation to climate and the many methods for managing water.

## Climate, Crop, and Season Extension

For market gardeners a great deal of thinking and effort have been expended lately on season extension, and rightly so. We want to produce as much as possible through all seasons, partly for marketing reasons, partly to satisfy human needs, our own included. Irrigation, hoophouses, low tunnels, and heated greenhouses, all promoted with pioneering innovations and trials by farmers like Eliot Coleman, have given us the ability to produce year-round throughout the continental United States and in parts of Canada. Unfortunately (or fortunately) the petroleum resources that go into many of those tools won't be around forever. But older methods of season extension are still in use around the world. And adapting to seasonality isn't all about growing tomatoes in December in any case.

The first and most important adaptation to changing seasonality, in fact, lies in choice of crop. Every market gardener who practices season extension knows this. There are crops you can grow in the dead of winter and those you cannot save through even the lightest frost. There are varieties adapted to winter cold or summer heat and those you want to avoid under those circumstances. There are winter wheats and spring wheats. Many fruit trees require a minimum number of chill-hours over the dormancy season (measured in hours spent below 42°F/5.6°C) in order to flower and fruit—and climate change is shifting favorable conditions out from under established local and regional cultivars. Thus drastic changes in local climate or the availability of water can call for a wholesale shift in crop choice and farming patterns. And the uncertain climate of the future means we have to be able to shift rapidly as conditions develop.[1]

Some of the most adaptive farmers of the older agriculture occupy arid lands with uncertain rainfall. Hopi blue corn and the Hopi lima bean, for example, are uniquely adapted to being planted deeply in sand dunes, where moisture is held by a layer of six to eight inches of dry sand. And, according to Gary Paul Nabhan, the Hopis and many other indigenous farmers are careful not to breed their crops "true to seed," thus intentionally preserving a variety of genetic traits in their seed stock. They may even sow "multiline mixtures of the same crop as a kind of insurance against disparate conditions." Russian grain farmers, European forage growers, and Latin American manioc planters do the same.[2] Even commercially produced Italian vegetable seed has greater genetic variability than typical American-produced seeds.

Locally developed and saved seeds reflect prevailing conditions and so are often favored by traditional agriculturalists today over commercial hybrids, which, in the case of corn and wheat, require heavy doses of purchased inputs to produce yields superior to traditional varieties. But prevailing conditions can change and are clearly changing now. Here, too, indigenous knowledge has something to teach us. As we saw in chapter 3, many indigenous farmers of the American Southwest and northern Mexico maintain multiple varieties of the same crop in anticipation that seasonal conditions will make a standard variety untenable. Mexico's Mountain Pima, for example, have several varieties of maize, including a quick-maturing corn planted when spring drought delays planting of the longer-season varieties. Other strains are adapted to early planting in a cool, wet spring but take five months to mature.[3] Peruvian and Bolivian potato farmers are famous for the wide variety of potatoes they grow, adapted not only to different uses but also to the many ecological niches these farmers have at their disposal, and thus to changing climate conditions as well.

Growing interest in seed saving today has moved from simple preservation to renewed efforts to develop varieties particularly suited to specific climates and locales. Small seed companies like Frank Morton's Wild Garden Seed make available genetic mixes for precisely this sort of development. And homesteader and plant breeder Carol Deppe not only produces seed adapted to the conditions she faces in Oregon's Willamette Valley, but provides clear advice on breeding varieties to meet specific requirements. Will Bonsall, farming in the very different conditions of Maine, has done the same.[4] Like our forebears from the birth of agriculture, we will need to develop an arsenal of varieties that can serve under different conditions if we are to be truly resilient.

Going further, we may have to reconsider our farming models altogether. Apparently the Little Ice Age, starting around 1300 CE, imperiled grain growing in Scandinavia and promoted the fortunes of rutabagas (high in vitamins C and A, by the way) and turnips. Both are still important staples in that region and well adapted to the moist, cool climates of Canada's Maritime Provinces and Alaska as well. The long recent drought in California, where I farm, has prompted many small, alternative farmers to think more seriously about taking advantage of our region's wet winters and turning to winter and spring grain production, as our forebears did through much of the state. Even so, erratic weather conditions threaten to defeat us. At the height of the drought in 2014, when virtually no rain had fallen before January, local grain pioneer Doug Mosel (founder of the Mendocino Grain Project) was prepared to throw in the towel if he lost his crop to continued dry weather. (Fortunately for all of us locally, enough rain came to make a crop, though the rains faltered again by April.) The latest climate data suggests continued drying for most of the western third of the United States. Strategies for resilience should include: adopting vegetable crops that could be dry-farmed or are drought-tolerant;

turning to more drought-hardy vines and tree crops; being prepared to abandon crops as water sources fail; and finding new ways to apply water conservatively (see below). The wetter, stormier, and somewhat warmer conditions being experienced in the Midwest and Northeast, meanwhile, will call for adaptations of their own, as will the loss of sufficient chill-hours to grow certain fruits in more southerly regions.[5]

Beyond the challenges that a drier or wetter climate mean for some of us is the continuing concern to practice season extension without the props of cheap plastic and fossil fuels. Here it might seem that traditional practices have little to teach us, that the advantages industrialization brought seriously increased possibilities for year-round food security. And it is true that traditional societies could face severe food shortages from time to time. It is also clear that modern agriculture has not put an end to famine, and in fact may well have exacerbated the conditions that cause it. Apart from that debate, the view that traditional agriculture and food provisioning simply came to a halt with cold weather and shorter days just doesn't stand up to examination. Nor was viable agriculture inevitably defeated by hot, arid climates.

Simple techniques for protecting vulnerable crops have probably always been a part of agriculture. A friend spent a fall and winter on a French farm thirty years ago working as an agricultural laborer. His job: covering and uncovering lettuces with a comfy bed of straw as protection against frosty nights. The Parisian market gardeners built on this tradition, using glass cloches to cover individual plants and glass-covered cold frames for seed germination and bedding plants. By the seventeenth century, but probably earlier, market gardeners in France and England had adopted the techniques of royal and aristocratic gardeners in using the same techniques to force plants, producing cucumbers nearly year-round, asparagus from October through February, and melons in May. Hot beds composed of a mix of raw and partially composted manure and topped with compost were then covered with cold frames, insulated by straw or more manure in the paths and protected at night with mats of rye straw. Depending on whether the underlying soil was wet (clay) or light, the hot bed might be dug in or raised above the level of the surrounding soil. And the compost that resulted from an exhausted hot bed would be used in turn to top the next. Interplanting and complex crop rotations established the basis for the French intensive gardening that soon spread to England and was propagated in this country by Alan Chadwick.[6]

Also in Europe, but widespread elsewhere, fruit trees, vines, and more heat-needy annuals were often planted against a south-facing wall of brick or stone to take advantage of the radiant heat retained into the night. From the sixteenth century onward, Northern Europe saw large acreages built out in fruit walls—south-facing walls against which fruit trees and vines were grown, sometimes espaliered against a wooden frame attached to the wall. Walls were built of cob or brick, sometimes topped with small roofs to protect trees and their fruit from rain. Later, glass frames were added as a lean-to against the wall for further protection and for raising oranges and even pineapples in northern climates. The Romans, Chinese, and Koreans had built similar structures, using mica or selenite (Rome), or oiled paper (China, Korea), but always employing a heat-absorbing wall behind. Modern greenhouses, by contrast, lose heat rapidly through glass or plastic on all four sides and must be heated at high costs in energy consumption.[7] In the Austrian Alps more tender fruit trees are planted against a south-facing wall, cliff, or boulder, and fruit trees are reportedly planted among boulders in parts of northern India. The boulders provide a microclimate and conserve moisture in the soil so the trees can thrive in otherwise inhospitable conditions.[8] Like the fruit walls, and requiring still less technological infrastructure, terracing can provide similar protection for a larger-scale, lower-cost planting. Many of the intricate terraces of the Andes are carved bowl-like into south-facing mountainsides.

In China today massive clay or rammed earth greenhouses provide season extension. They are covered with plastic sheeting, which can be covered at night with a textile mat much like the Parisian hot beds. Also in China, low earthen walls erected around small plots protect seedlings for an early start. These, too, are often covered at night.[9] A common practice throughout the world is to surround plants with heat-retaining rocks to protect against nighttime chill. In arid climates this "rock mulch" also serves to conserve moisture and moderate soil temperatures throughout the day. Likewise in arid regions, seeds are planted in furrows rather than on ridges or beds, giving protection from the sun in the early stages of germination as well as privileged access to moisture. Simple shade structures built of forked sticks and mats are used in Mexico to protect seedbeds and seedlings from the sun and maintain soil moisture in the hot, dry months before the first summer rains. Though even traditional farmers in Mexico dispute the utility of the widespread practice of hilling up soil around corn plants to prevent lodging (toppling over in windstorms), it seems clear that doing so preserves moisture near the roots and provides soil for additional rooting in a region where adequate rainfall is always uncertain.[10]

Many crops enjoy the protection of other plants and under certain conditions can only be grown with that protection. This is the principle embodied in the permaculture practice of multistory plantings. Shade-loving annuals, or those that require cooler conditions, can be grown in a hot climate and out of season with the protection of upper-story plants like bushes or fruit trees, which in turn may depend upon a higher canopy of nut-bearing trees. While many crops have been bred for full sun over the last century or so, and produce more under those conditions, these varieties do not necessarily have the quality of more traditional shade-grown varieties. Many coffee producers and coffee buyers thus prefer shade-grown varieties to the newer cultivars. More important from our point of view, the newer varieties cannot be grown without adequate sun, limiting where they can be useful. Other examples of multistory systems come from West Africa, where farmers nurse young cacao plants under plantain and interplant with cowpeas, cocoyam, taro, cassava, and yams.[11]

Intercropping offers other benefits, often including higher overall yields per acre and a ready form of "crop insurance." More drought-tolerant crops may thrive where water-hungry ones falter, and vice versa. As Paul Richards puts it, "Where a farmer in Europe borrows from the bank to tide him over a bad year and prepares for better results next time, the West African farmer may not survive to try again." Intercropping can also even out labor demands, allowing farmers to weed multiple crops at once while planting over a period of time. It is also used to provide a wide variety of foodstuffs to farmers who depend on farming for much of their own sustenance. Richards found up to sixty species planted to a farm in the forest zone, though twenty to thirty was the norm; in the savanna farms, on drier soil, farmers planted ten to fifteen species together.[12]

Storage is another ancient technique for extending the season. Most root crops are biennials, the roots dedicated to storing up nutrients for a burst of seed production during the next season. That's why they have traditionally been popular for long-term storage, either in the ground or in simple storage arrangements aboveground. Some roots have been grown specifically for winter forcing. Witloof, or Belgian chicory, is a winter delicacy in much of Europe (and occasionally in this country). The root is grown over a normal season, then trimmed of green leaves and packed upright in sand for the winter. When the sand is moistened, the chicory begins to sprout, producing a tight, blanched head used in salads or split and roasted.[13] Semi-nomadic peoples in Southeast Asia and the Amazon often planted tubers, such as sweet potatoes, cassava, manioc, or white potatoes, in untended gardens, leaving them for harvesting in part or whole on the group's return or as need arose. And many winter market gardeners know that the best place to "store" winter vegetables is in the garden. Eliot Coleman was not the first to winter over carrots for the intense sweetness of the frost-touched root; cabbage, kale, brussels sprouts, and collards are well known to sweeten with frost.

We are familiar, in fact, with a number of conventional storage crops. But most of us do not store them because we think we lack the facilities to do so. Probably the biggest exceptions are garlic and winter squashes. Many small farmers save garlic seed, which can be stored from harvest in June until planting in October without complicated cold storage. Winter squash can be cured under conditions most of us can arrange and stored through much of the winter at 50 or 60°F (10 to 15°C), becoming sweeter as it ages. But other crops require more exacting conditions, or so we think, and thus equipment that is out of the reach of the ordinary small farmer. Nevertheless, traditional agriculturalists often stored a variety of crops without the help of electricity or rigid foam insulation. My great-aunt inherited a double root cellar along with the six-room log cabin she and her husband moved into in rural Montana in 1912. Root cellars were common even in urban areas in the United States until the 1950s. This one had a warmer forechamber, used primarily for storing Aunt Anna's milk as the cream rose before she churned the butter. The rear chamber was deeper in the earth and better protected from frost, and this was where she stored canned goods, potatoes, squash, and roots. The cellar was constructed of wood timbers and planks, dug into the earth, and mounded with dirt. Other cellars in both rural and urban areas were built of brick as protection against rodents but also dug into the backyard and bermed.

These cellars probably go back very far in human history, as do various sorts of granaries. But even where permanent construction was avoided, traditional societies stored food successfully. In California many indigenous groups buried fall-harvested acorns in streams, where passing water would leach the tannic acid out of them, making them sweet and edible when the group returned at the end of the winter. Roots were stored in tightly woven baskets filled with sand or a light, moist soil. Euro-American families have used the same technique for several centuries. Another traditional technique is the construction of "clamps," stacks of vegetables covered in straw and mounded with dirt for storage through the winter. Clamping was once a widespread form of winter storage for both farmers and homesteaders. And many cultures have used pottery urns sealed and buried in the ground to preserve foods of all kinds.[14]

Fermentation is a specialized form of food preservation—though to Westerners the ferments enjoyed by some cultures just taste rotten. But simple fermentation, whatever the flavor outcome, protects food from contamination by dangerous organisms while favoring those beneficial microbes that are good for the human stomach. In Korea kimchi is prepared on a large scale at harvesttime, combining cabbage, salt, chilies, and other vegetables. The prepared cabbage is traditionally placed in large earthenware pots that are buried underground, where cooler soil temperatures moderate the ferment and where the product can be kept for months. In Germany and Austria, similarly, whole cabbages or shredded cabbage are stored in barrels in a brine, to be used a bit at a time over the rest of the year. Pickles of all sorts used to be prepared in similar ways. Cheesemaking is perhaps the most familiar of these techniques for preserving and enriching a seasonal product and storing it for long periods of time. These are very old practices that are found throughout the world.[15]

Many of us would benefit right now from adopting some of the cultural practices we've seen here, and some, like small-scale fermentation, are already popular. Older technologies for extending the season will be useful to farmers who want to wean themselves from our current reliance on plastic. These techniques are also examples of what John Michael Greer calls the trailing-edge technologies that we are likely to employ as fossil fuels falter and cheap energy and industrial products become increasingly scarce. And the salvage economy that may also follow is already familiar to many of us who rely on the discards of industrial society for building and machinery.[16] To take just two examples: The single-pane windows made obsolete by new building regulations can be repurposed as cold frames and glass-faced fruit walls over the coming years. And simple resources available right now can provide storage facilities, like the root cellar and other traditional storage techniques, independent of fossil fuels and the grid.

## Water: The Indispensable Resource

In 2012 the US State Department published *Global Water Security: The Intelligence Community Assessment*. The collective judgment of the US intelligence establishment was striking for its willingness to attribute growing shortages of water to climate change. But more important was the high confidence of the intelligence agencies that by 2030 global demand for fresh water would exceed supply by 40 percent. Given the authors, it is not surprising they saw this as a national security issue.[17] The report is not the first to call attention to faltering water supplies. Whether the causes are declining levels of snowmelt or falling aquifers, farmers around the world are already feeling the impact of water scarcity. At the same time, there are areas of the world, including the American Midwest and Northeast, that are experiencing unprecedented storms, with rainfall that threatens crops and infrastructure. The historic downpour that accompanied Hurricane Harvey in the fall of 2017 was record breaking. Wise water management cannot stave off truly devastating drought or flooding, but it can make us more resilient in the face of increasingly uncertain weather conditions. In this respect, too, we may have much to learn from traditional farming cultures.

Farmers through the centuries and around the world have confronted wildly diverse challenges in managing water. High water tables and low; seasonal uncertainty and predictable flooding;

raging rivers, slow seeps, intermittent springs, and distant sources present diverse challenges to water-hungry farmers. And farmers' responses have been, if anything, more diverse. From elaborate drainage and channeling schemes, to multiple strategies for storing and dispersing water, to unexpected ways to conserve and ration water, farmers have devised thousands of ways to manage water supplies, whether scarce, abundant, or unpredictable. As the ready energy of fossil fuels draws down, as larger-scale irrigation schemes face the new challenges this entails, and as we adapt to more uncertain weather and precipitation patterns, we will have to find new ways to provide water to our farming operations.

We have already explored some of the ways farmers managed water in the previous chapter where we looked at use of floodwaters for fertility and terracing as a form of soil conservation. These same techniques are also forms of water management, putting moisture where it is most needed and conserving it for plants. Leveling fields, bounding them with terraced steps or low mud barriers, and holding water in paddies and ponds are all means of conserving moisture while putting water to use to enhance fertility. But historically farmers have also faced the flip side of this coin, contending with sodden and flooded fields at times when they needed to plant. The history of drainage and its many techniques is also a story of water management, with the highly productive farms of the Netherlands as the exemplary model that inspired generations of similar efforts in Europe, England, and North America. But drainage has also created enormous ecological disruption and destruction, and there would seem to be few circumstances today where large-scale drainage, or even smaller efforts, would be justified. Our remaining wetlands should stay wetlands and, if anything, should be "cultivated" lightly for the unique plants that only they can produce, as our indigenous forebears did. Already, former wetlands are being restored on a small scale.

Small-scale drainage, nevertheless, is important to farmers on heavier soils in temperate zones, who need to plant in time to make a crop. Channeling and buried tiles are very old methods of draining off enough water to allow a field to dry in time for planting. But farmers in some zones face a relatively high water table year-round. Wet soils can be left to meadows and hayed in the dry season or minimally drained for farming, and drainage may also serve for irrigation. Fields in parts of Mexico and Central America are drained by networks of canals, emptying into streams or lakes in such a way that the water table is kept low enough to permit cropping, while the drainage water provides irrigation for seedlings and those plants whose roots do not reach the depth where capillary action provides subirrigation. Gene Wilken notes that "the objective is to achieve relatively rapid transfers of water from *zanjas* [ditches, canals] into plot substrates during dry seasons, and reverse flows from plots to *zanjas* during rainy periods to remove excess moisture. Flows are substantial." And this generally means that plots must be small enough to permit effective drainage and moistening.[18]

While Mexican farmers in wetter areas do not manage the canals to ensure controlled subirrigation and prefer to irrigate from the surface, they certainly benefit from the high water table. The chinampas—sometimes referred to as "floating gardens" of the Valley of Mexico—depend upon the lagoon that surrounds them for subsurface irrigation. Fields bordering Lake Atitlan and swamps elsewhere in Guatemala probably depend on subsurface moisture. And both sunken fields and raised beds uncovered by archaeologists from central Mexico to the coast of Peru may be explained as part of schemes for harvesting water below the soil's surface. Such soils are vulnerable to salinization, which traditional farmers combat by the labor-intensive practices of adding sand to the soil to enhance drainage or scraping salt from the surface.[19] Chinese farmers avoid the problem of salinization by frequent emptying and turning of paddy soil. Rice paddies are drained after harvest and long beds or mounds built for a second and third crop. The soil is still quite wet in most cases, and the furrows hold water, so plants eventually enjoy water through capillary action from the still-high water table.

In coastal West Africa farmers use estuarine wetlands, allowing seasonal floods to wash out salts, then planting wet rice as the water level recedes. Archeologists speculate that such areas may have been the birthplace of grain agriculture in Mesopotamia. Such "flood retreat" planting continues today in many parts of the world. West African farmers carry the process further, cultivating swamps in stepped terraces with low walls formed on contour. Farmers carefully select rice varieties according to the moisture conditions of the soils they are working with. An individual farmer may work with as many as ten varieties of rice, his village employing up to thirty-five. After the rice harvest they plant wheat, tobacco, and vegetables to take advantage of the remaining moisture in the soils. Similarly, terraces are farmed on the lower levels of river valleys, capturing the seepage from higher up. Early-season crops rely on the silty soils of the lower hillsides. Quick-ripening varieties of rice or early yams complement the main, later-season crop farmed on the uplands, and may take advantage of higher prices early in the season.

In more extensive floodplains, complex systems have been devised to make use of the seasonally variable water resource. In the inland delta of Nigeria, farmers plant rice on shallowly flooding bottomland, and cotton on the bars. Dikes are built to protect the rice from undue flooding, while channel irrigation supplies water to the cotton. During the dry season, as in swamp cultivation, vegetables, tobacco, and wheat replace the rice, drawing on residual moisture in the soil and, sometimes, bucket irrigation.[20]

The principal challenge that traditional farmers have faced is getting sufficient water to their crops in a timely manner. Irrigation, as we saw, is probably as old as agriculture, certainly in the Middle East. The means farmers have devised for meeting the challenge are legion. Egypt's annual flood irrigation required flood control, accomplished through linked series of basins to hold and release the floodwaters of the Nile. Mesopotamia, like much of the world, utilized permanent canals, maintained with heavy labor investments over centuries. In canal systems silt must be periodically removed, and it was often spread by hand or oxen over neighboring fields, sometimes carted long distances to fertilize more distant plots. Extensive canal systems require management, suggesting to Marx and later Wittfogel the need for a complex administrative structure, prompting the rise of so-called Oriental despotism.[21] But most irrigation systems, we now know, have been governed by the users themselves, joined together in democratic water associations. The most famous examples are the acequias of New Mexico, but the Spanish *sociedades de agua* are still more ancient. We will look at traditional governance arrangements in chapter 8.

For most of these systems, however, there are two requisites beyond irrigation canals. First, the water must be available when crops need it most, which sometimes means some sort of water storage arrangement. And second, the water must be lifted from the canal to the fields somehow. Traditional farmers have been inventive on both scores. Not all farming societies rely on stored water, of course. Summer rains may set the terms for agriculture, determining what and when to sow. In many parts of the world, farmers have only (!) to face the uncertainty of rainfall—whether it will come on time, be sufficient over the life of the crop, or wipe out a crop with excess water. In such places, too, irrigation and even water storage have been practiced; or farmers cluster around a lake or river that can provide a steady supply of needed water, or take advantage, as we saw, of seasonally high water tables in bottomland, wetlands, or estuaries. But the greatest challenges lie where water is scarce and unseasonable from the point of view of the farmer.

The modern strategy of irrigation based on large-scale damming and storage of water to feed canals was not available to traditional farmers, who lacked the capacity to build those dams. As we saw in the last chapter, small-scale check dams and temporary structures to channel river flows could spread and help retain moisture over time.

But they could not provide permanent sources of irrigation water. Nevertheless, some cultures, notably in India and the Middle East, developed medium-sized storage systems to capture rainwater or intermittent surface water and make it available for irrigation uses at the appropriate times of the year.

In drier parts of India like Rajasthan, bermed tanks of stone or concrete were constructed upstream from fields and on the edges of towns to capture monsoon water. Surrounded by a round catchment area, carefully cleaned and graded before the rains came, the underground tank filled with the rains to provide water in the dry months. Smaller tanks served drinking water needs. Larger ones, constructed by the village or by a rich villager on behalf of the village, provided irrigation water. When irrigation was needed, water was channeled from these tanks to flood-irrigate downstream fields in sequence. Tanks and canals were managed by a single village, as was the distribution of water among farmers. With the introduction of Green Revolution crops and pump irrigation, many of these systems fell into disuse. Poorer farmers who could not afford the inputs and pumped water struggled to survive. As water tables have dropped due to overpumping, the older storage and irrigation schemes have been revived by village committees.

Many Indian and Middle Eastern towns and cities have elaborate belowground cisterns, some of them open, others covered, where drinking water and limited irrigation water is stored. Some are as large as a palace and several stories deep. The open canals that sometimes feed them can be a source of contamination in some cases, but many are fed through closed canals. Some are fed by underground springs or by *qanats*, underground tunnels tapping springs or aquifers on higher ground. Whether rainwater-fed or dependent on subterranean water sources, cisterns provide a reliable, if limited, source of water in arid climates. Cisterns dating back to the fourth millennium BCE have been found in Lebanon.

In the Yucatan much of the water upon which Mayans have relied for centuries is available in *cenotes*, underground caves where rainwater naturally accumulated. Regarded as sacred spots by the Mayans, cenotes were carefully tended and often hidden from outsiders. Where local geology did not provide cenotes, Mayans made use of natural shallow depressions or basins from which building material had been removed, often lining them with lime to prevent seepage. And in northeastern Yucatan, bottle-shaped cisterns, called *chultuns*, averaging seventy-five hundred gallons in capacity, were constructed under houses or ceremonial buildings to supply local water needs. Rainwater was channeled from roofs and paved areas to keep chultuns filled.[22]

Besides feeding cisterns, qanats are a unique and important method of irrigation in dry lands. Qanats (called *galerias* in Mexico) have been recorded across the Middle East, going back to the seventh century BCE. Brought to Spain by the Muslims, the technology is also found today in Mexico, Peru, and Chile, though this form of horizontal irrigation may predate the Spanish conquest.[23] The tunnel taps a belowground water source upgrade from users' fields, bringing water to temporary storage or to canals that serve the fields. Well diggers lay out a line of dry wells that will provide air vents for diggers and an outlet for the soil that is excavated. They dig from well to well, finally reaching the water table. The tunnel is constructed on a gentle grade and lined as needed to prevent cave-ins. It must be maintained over time, as silt or debris from earthquakes may clog the waterway, so the dry wells along the way are left open. The qanat may be as little as a few hundred feet or many miles long. Qanats remained a primary source of irrigation in Iran, Syria, and elsewhere into the 1970s and were still important in places like the Tehuacan Valley of Mexico in the 1980s, where they are cheaper to construct and maintain than deep wells and pump irrigation.[24]

Where the water table is close to the surface, or irrigation canals bring water to the fields, farmers still have to get the water to the plants. In arid regions where water is scarce, flood irrigation is not desirable. Moving water is difficult and tedious, but traditional farmers have devised all sorts of mechanisms for doing so. The simplest is hand irrigation. Using shallow, hand-dug wells in their fields, farmers in parts of Mexico haul up water in specially shaped pots or plastic jugs. In Oaxaca and Tlaxcala, Mexico, larger crops are planted in shallow basins, smaller ones in small, ridged plots. Farmers carry water across the field, carefully filling each basin or plot. In some cases fields are just big enough for one worker to water in a day, and wells are placed to maximize efficiency in watering. In the *chinampas*, water from canals is pulled up in buckets and splashed on plots. The narrow length of plots and the ready availability of water enables farmers to manage shallow-rooted crops and seedlings whose root systems are too small to benefit from the high water table.

Where the water level in channels is closer to the surface of fields, farmers around the world have resorted to splash irrigation, lifting the water in long-handled shovels and splashing it on the surface of plots. Farmers construct narrow plots along multifingered irrigation channels to make this approach work. In Korea, and occasionally in China, F. H. King found farmers using a swinging scoop to lift water a few feet from canal to field. Variations on the Egyptian *shadoof*, a bucket balanced on a long staff by a counterweight and designed to raise water several feet, are found throughout the world. But the most common methods that King encountered in China and Japan were variations on the waterwheel. Long wooden chain pumps carried water up steep embankments, using human power through foot pedals or animal power hitched to wooden-geared cranks. Waterwheels raised water to the canal edge, dumping it directly into paddies or filling narrow channels to fields. In Japan workers trod the circumference of the wheel in a useful version of our gym treadmill. Waterwheels moved more water than container or splash irrigation, but all these methods consumed lots of human labor, for which only productive fields could compensate.

Water was also conveyed in tubes by gravity in many parts of the world. The Romans famously used lead to fashion water pipes for domestic consumption. High concentrations of calcium in the water supply meant the lead surface was quickly coated, preventing the worst forms of lead contamination. In the Middle Ages, hollowed-out logs fastened together with wooden pegs and animal fat or tar were used for conveying water for domestic use and irrigation; such pipes were being laid in the United States as late as the Second World War.[25] Bamboo has been used elsewhere. In northeastern India farmers tap springs in the hills and bring water as much as several miles to fields and homes on split-bamboo structures. The water is delivered to plants in drips through holes in the final segment of the channel. Irrigation systems are rebuilt every two to three years and maintained by cooperatives.[26] Elsewhere in India both hollowed-out logs and bamboo are used to transport water from streams and springs to villages.

Among all these ingenious practices, the most applicable today and in the near future are likely those surrounding water storage, conservation, and wise water use. Many of the traditional water conservation strategies mentioned here have been explored in Brad Lancaster's two-volume *Rainwater Harvesting* and Gary Paul Nabhan's *Growing Food in a Hotter, Drier Land*. The waffle gardens of the Southwest, like the small ridged basins relied on in arid areas of Mexico, suggest alternative planting schemes where water is scarce and can be delivered precisely to plants or through controlled runoff. In wetter areas the techniques of primitive subirrigation reviewed here can be appropriate, taking advantage of high water tables while providing adequate drainage through raised beds or mounds. And these can be important, at least on a small scale, where winter rain and snow usually prevent early sowing of crops.

As for storage, while poly tanks are relatively cheap and readily available, they are not the most durable option, and they can suffer from overheating and contamination in hot climates where they are not shaded. Ferro-cement tanks can be constructed in large sizes to store pumped water; the same technology can be used to build underground cisterns to hold rainwater from rooftops or constructed catchment areas.[27] Paddy culture has been widely adapted, though we have only recently begun to appreciate the biology and chemistry of submerged soils. As the ancient history of this approach suggests, we can find simpler and less expensive ways to construct temporary and even permanent ponds than the artificial fabric-lined structures favored today. Many of the alternatives have been studied and tested in the permaculture community, from muck- and manure-lined ponds to crater gardens that double as seasonal, open cisterns.[28]

The long haul will bring hotter, drier conditions to some of us, and wetter, less predictable weather to others. Everywhere we will need to adapt, as farmers always have, to changing conditions. And we will need to be able to farm more marginal land well, regenerating soils where they have been degraded, dealing with water management issues that may be unfamiliar today, and finding ways to cope with changes in seasonal response by plants. The enormous repertoire of techniques developed by farmers over the centuries, some of them recovering from disastrous farming practices in their own pasts, are a resource we should be looking to as we shape our own adaptations today and in the future. And those traditional farmers have still more to teach us about the "whole farm" that can sustain our livelihoods for the long haul. The next chapter takes up that theme before we turn, in chapter 8, to the social ecosystem on which we also depend.

## Chapter 7

# Woodlands and Wastes

*The white man sure ruined this country. It's turned back to wilderness.*

—James Rust, Southern Sierra Miwok elder[1]

Modern scientific forestry was born in Prussia in the late eighteenth century. Its radical simplification of forest management became the gold standard for enlightened state officials around the world, inspiring Gifford Pinchot, who decisively reshaped US forestry policy under Theodore Roosevelt and laid the groundwork for the present system of management. For Prussian forest managers the central concern was the sustainable yield of timber that could be extracted from state forests annually. Missing from this vision, as James Scott comments, "were all those trees, bushes, and plants holding little or no potential for state revenue. Missing as well were all those parts of trees . . . which might have been useful to the population[:] . . . foliage and its uses as fodder and thatch; fruits, as food for people and domestic animals; twigs and branches, as bedding, fencing, hop poles, and kindling; bark and roots, for making medicines and for tanning; sap, for making resins; and so forth."[2]

State foresters, in both Europe and the United States, were at pains to regulate large-game hunting, but ignored when they did not forbid "the vast, complex, and negotiated social uses of the forest for hunting and gathering, pasturage, fishing, charcoal making, trapping, and collecting food and valuable minerals as well as the forest's significance for magic, worship, refuge, and so on."[3] The principles of state forestry were only one version of the closing of the commons to traditional uses by ordinary people. The extension of the principles of state forestry to state and federal lands in the United States, like the enclosure of the commons in England and similar measures throughout the European colonies, stripped local users of legal access to much of what once made up subsistence. The loss has been incalculable.

Traditional agriculturalists in so-called tribal societies generally combined horticulture with hunting and gathering. Kat Anderson summarizes the situation among California's Native Americans:

> Every plant community in California—from beach dune to evergreen forest and pinyon-juniper woodland—was visited by gatherers, and within those communities every type of plant life form was gathered. . . . Every kind of plant part—underground bulbs, rhizomes, and roots; the oozing resin of trees; the sweet nectar of certain flowers; stems, bark, branches, shoots, leaves, thorns, flowers, seed pods, seeds, seed plumes, and cones—found a use.

And as Anderson shows in an extensive survey, Native Americans cultivated the wild through a whole series of strategies, aimed at sustainable harvests of a wide diversity of species of value.

From controlled burns, to coppicing, to selective harvest, to scattering seed, to tillage, so-called hunter-gatherers tended the wild, significantly altering, and often enhancing, natural ecosystems.[4] Uses were carefully governed, as they were also governed in peasant communities around the world. Specific hunting, fishing, or trapping sites might be allotted to particular hunters, though rights to use were rarely exclusive within the group. Locations where tubers and roots could be dug, or reeds or willow wands harvested, could be the preserve of one or another family. Familiar mushroom stands, berry brambles, camas meadows, or acorn oaks might be widely shared—and carefully wildcrafted—or reserved for individuals or families, with gleaning rights for other members of the community. Wild herbs, poles and saplings for building, firewood, fibrous plants for cordage and weaving—all these were available from the land base of traditional agriculturalists as well as so-called hunter-gatherers.

Among the privileges that peasants enjoyed under feudal arrangements around the world were secure access to land for cropping, common pasture for feeding a prescribed number of animals, firewood and building materials from forest, waste, and wetlands, fish and small game such as rabbit or squirrel, and foraging rights for herbs, mushrooms, and medicinal plants. Cutting of reeds or willow wands for basketry and related crafts was often part of the package. Where clay was available it could be had for pot-making by the family or a village potter allotted the task of making pots for everyone in exchange for other goods. Thatch for roofing would be gathered from field straw, reeds, sedge, heather, or palm fronds set aside for families or a village thatcher. In developed feudal societies, specific rights to all these resources would be carefully enumerated as part of the obligations owed to a household or peasant community by the feudal lord. And even as feudal ties weakened and peasants and others claimed property rights, traditional rights to the fruit of the commons persisted into the modern era.

In both tribal and peasant societies, there was no "tragedy of the commons" so long as the natural resources available to the group were communally managed. As we will see in more detail in chapter 8, such management was most often carried on in democratic councils of shareholders. And even where feudal lords had authority, custom and the guardians of custom held sway over the commons. When customary arrangements were abused or disregarded, peasants had recourse to manor courts and sometimes the royal court. The disputes that issued in the Magna Carta of 1215 originated in conflicts between nobles and the throne over jurisdiction, with the king defending the rights of peasants, and these disputes continued apace, up to the beheading of Charles I in 1649. But when more powerful individuals and families gained exclusive power over the distribution of resources, exploitation often followed, as officials increasingly distant from actual use extracted more and more for the benefit of the few. The real "tragedy of the commons," Simon Fairlie shows, was private (and sometimes public) appropriation for profitable use.[5]

Successful management of specialized commons, in fact, has persisted for a thousand years in some well-documented cases. In the Alps, villages have carefully apportioned out village pastures among householders for centuries, with herd size controlled to optimize the reproduction of grazing lands. In Spain, irrigation systems reaching back almost a thousand years have persisted under the management of users. We'll look at how these common resources were managed in chapter 8.

To what extent traditional agriculturalists depended upon some sort of commons beyond their own farming efforts has varied enormously from society to society. But everywhere the natural resource base beyond crop and pasture land has provided at least an important supplement to agriculture. Up until recently in the United States, where the practice of village-owned commons largely fell out of use after the seventeenth century even in New England, farm families depended upon a whole farm that included woodlands, streams, and marginal land for supplements to their crop and livestock production. Those supplements included firewood most commonly, but also grazing land for cattle and sheep, browse for goats, as well as fish and game, herbs, and mushrooms among those who knew what to look for. They also included, we now know (many knew then, if in other terms), such ecosystem services as habitat for beneficials (as well as pests and predators), buffer zones from storm, heat, and, in places, wildfire, and water retention or drainage.

As private property erected visible and invisible fences between neighbors, and the state tightened control over state and federally owned lands, access to many of these resources fell out of the reach of many rural Americans. The earliest state and national forests, for example, were generally established on land already occupied by Native Americans, *métis* (families of mixed French and Native American heritage), and squatters of all races.[6] Everywhere state and federal authorities have pursued the expansion of the public land systems through the extinction of preexisting rights, either by means of outright expropriation (applied most often to Native Americans and poor whites) or by purchase when property comes up for sale. While public authorities have recognized more and more uses, from hunting to mushroom gathering, governance is always in the hands of the bureaus, with little recourse for local users when policies change in Washington or the state capital. And for some, like Blackfoot Indians with traditional claims to hunting and sacred sites in Glacier National Park, exclusion seems to be permanent. Where private property dominates, local custom can be equally exclusionary, as in much of the American West, where landowners often protect their claims to exclusive use at gunpoint.

Changes in farming practices and everyday culture have also undermined traditional uses of rural land. Focus on production first, pioneered in California's rich valleys already in the nineteenth century and implicit in grain farming for export in the post–Civil War Midwest, became the dominant mode of farming by the mid-twentieth century. Agricultural chemicals brought a boost to production, raising the persistent specter of overproduction and falling prices. Farmers expanded in response, consuming more and more of the available land for cropping. Encouraged by the USDA in the 1970s to farm fencerow-to-fencerow, American farmers abandoned the diversified family farm of the past, as we have seen, pulling down barns and outbuildings, cutting woodlots, and abandoning kitchen gardens. Marginal land was either farmed to its ruin or abandoned to weeds. The "whole farm" on which farm families had depended for subsistence was exchanged for a government check and a job in town. And in the meantime, the farm family had acquired tastes for purchased products that made the rat race of production seem inevitable in an economic system rigged against farming.

## Managing Woodlands, Utilizing Wastes

To step into Ben Law's *Woodland Way* is to encounter a world once familiar to Europeans but almost unknown in the United States. Law is one of many in Great Britain reviving the ancient arts of coppicing, charcoal making, wood crafts, and sustainable woodland management.[7] The new woodland workers are reclaiming traditions of forest- and woodland use that date back centuries, if not millennia, and that have parallels around the world. Much of Law's work is with coppiced woods, many of them now overgrown and in need of restoration. Coppicing is the practice of cutting trees and shrubs back to the stump to stimulate regrowth, generally of straight new shoots that will be harvested in anything from one year to fifteen or more. Most trees and many shrubs can be coppiced again and again, and the stimulation to their root systems actually extends their life beyond what would be expected of a naturally growing tree. The practice has been found throughout the world and is apparently very ancient. Pollarding, or cutting higher on the trunk where browsing animals cannot reach the new shoots, is also an ancient practice.

Native Americans often coppiced to provide smooth shoots for basket-making. Pomo Indians coppiced or pruned narrow-leafed willow to stimulate the growth of long lateral roots used in some of their larger baskets. Indians cut or burned a wide variety of shrubs and small trees to stimulate the growth of straight shoots, which might be stripped of their leaves for a year or two to prevent branching. The smaller shoots were used for baskets or arrows, stouter ones for fish spears and bows, and others for digging sticks and other utensils. Coppiced wood could also provide poles and stakes for building. And pollarded oaks yielded more acorns on the multiple branches that sprouted from the trunk.

In contemporary England woodworkers use coppiced wood for stakes, poles, materials for fences and panels, furniture, canes, and utensils. Charcoal producers burn older wood and wood that is misshapen for charcoal, a local substitute for a commodity that today is often shipped from as far away as Indonesia. Charcoal burners may also provide biochar for gardeners and charcoal fines for filters. Coppice plantations are generally small (less than three acres) and of a single species. But they may be interspersed with coppices of other species and more diverse woodlands used for timber harvest. Managers may cycle coppiced trees through shorter and longer cutting cycles, with some stumps, or stools, allowed to grow up a single standard for eventual timber harvest, while others nearby continue to be harvested for multiple shoots over shorter periods of time.

Mixed hardwood forests were (and are) also harvested sustainably in many traditional settings. Saplings may be thinned for poles and stakes. Underbrush could be used for woven work, including mud-and-wattle walls, or harvested for medicinal purposes. Bark, roots, and flowering plants can all be harvested for medicinal uses, dyes, and fiber. Sap is tapped for syrup and wines. What cannot be incorporated into some local use will often be cleared to open up the forest for game and to prevent buildup of flammable material.

Apart from wood products and the wildlife of woodlands, trees have been important to humankind for fruit and nuts since long before the advent of settled agriculture. The tree crops that USDA geographer J. Russell Smith researched over a lifetime of study were often central to the livelihoods of traditional farmers. In the 1920s Smith found prosperous Corsican peasant communities across the forested mountainsides of their Mediterranean island at two thousand feet in elevation. Their principal crop was chestnuts, from which they made a flour for bread and cakes. Chestnuts also fed their horses, dairy goats, and pigs. Chestnut-fed hogs were (and are) a specialty of many of the more mountainous regions of Southern Europe, prized by consumers and chefs alike. The chestnuts were all grafted, and reportedly some grafted trees were as much as a thousand years old. Chestnut cultivation apparently reaches back more than a millennium. The chestnut forests are relatively open, with new trees planted next to their dying elders to take their place in the sun when the older tree comes down. Peasants scythed the forest prior to harvest; one can imagine that it was once burned. In Portugal similarly, where chestnuts are grown more for wood, the ilex oak matures early and provides a reliable acorn mast for pigs. Chestnuts and oaks are just two of the tree crops that Smith hoped would be the basis for a perennial future agriculture, a hope echoed today by Wes Jackson and Wendell Berry and pursued by Mark Shepard and others growing chestnuts, hazelnuts, and berries and incorporating managed grazing.[8]

Traditional orchards were often "pasture-orchards" with large standard-sized trees set in extensive meadows, a model Mark Shepard advocates today for a permaculture farm. Examples can be found among both domesticated fruit trees and wild species valuable for their fruit or nuts. In the latter case the orchard was maintained by local gatherers; in the Amazon, as we saw, Native Americans actively transplanted and bred tree species for human use. Where Europeans historically used metal tools and livestock for clearing and maintaining woodlands, Native Americans and other cultures used fire. The first Europeans to arrive in what is now the United States described forests so open that one could drive a carriage through them. Periodic burning to increase habitat for game, rejuvenate valuable plants, reduce insect predation around favored species, and clear out the duff that could lead to catastrophic forest fires transformed the forestlands in all but the most remote and mountainous areas of the continent. There nature did the work with lightning-set fires. Today that work is accomplished by rare sustainable harvesting practices on lands outside the purview of state and federal forest authorities and, increasingly, by enlightened forest management practices employing controlled burns.[9]

The meadows and prairies that Europeans encountered in the New World were likewise maintained through regular, conscious fire management, preventing the encroachment of trees, brush, and chaparral, enlarging the area of grassland, and encouraging grasses and valuable forbs by clearing away dead matter and providing fresh fertility. It may well be that swidden agriculture grew out of such practices. It also appears that Europeans used fire in a similar fashion even after the introduction of the more labor-intensive metal tools.

Traditional users of the waste and woodland surrounding cultivated areas managed them in a variety of other ways not all that different from cultivation. Selective harvesting of a wide range of plants, from edible bulbs to mushrooms, encouraged reproduction and, sometimes, enlargement of a stand. Digging up wild bulbs, rhizomes, and tubers could have the effect of tilling, allowing remaining plants to grow larger; this was often the intention of the diggers. Many groups spread wild seeds that were of value to them, enlarging their habitat or making for denser plantings. Pruning, both for use and as a kind of plant care, stimulated regrowth and removed wood that inhibited optimal growth. Many peoples pruned wild as well as domesticated trees, knocking down dead branches and tips, burning off old palm fronds, or stripping suckers and opportunistic growth. Cattail, tule, and reed beds would also be burned to clear out dead foliage and make room for new growth.

Forests and savannas were favored sites for human habitation and perhaps the development of the earliest agriculture, but other landscapes were also valuable and sometimes intensively used and managed for their products. As we saw in the last chapter, floodplains could be cultivated as waters receded, and Mesopotamian agriculture surely began in the rich alluvial plain of the Tigris and Euphrates Rivers. In addition, reeds for mats, baskets, thatch, and even boats were harvested by lakeside and riverine peoples in sustainable fashion. Sturdier water plants like cattail and tule were cut for food and fiber. And everywhere fish, whether stocked in ponds or found in the wild, were an important supplement to diets.

In England the fens were places for refuge from the manorial system where free peasants relied on fishing, furtive cropping, and marsh resources of all kinds. On seacoasts, mussels, clams, abalone, and other shellfish were part of a regular diet, or sought out at appropriate times of the year by even relatively distant groups. Seaweed, too, was harvested for consumption in many parts of the world. And those who could do so managed salt basins for harvesting salt, which was traded far and wide. Salt deposits elsewhere were also visited and exploited throughout the world.

Clay deposits are widespread, in stream bottoms, around hillside springs, and exposed where roads cut below the surface. Clay construction is equally widespread, with clay-laden soils forming the basis for cob building, and purer clays used for adobe, brick, and plaster for ovens as well as for houses and outbuildings. Finer clays, of course, are the basis of the potter's art, and the pots have most often been fired in a clay oven. Clay homes and floors are plastered with a mixture of lime or manure and clay and often colored with earths dug for their coloring alone. They might also be painted with earth- or plant-based pigments.

Other deposits, from sand and gravel to gypsum and limestone, have been exploited from time immemorial for small-scale building, cement construction, and soil amendments. Lime makes a highly durable plaster, accounting for the ability of cob structures in the wet southwest of England to stand for centuries. Its properties have been known for more than seven thousand years, despite the difficulties of making it. Limestone, chalk, or even seashells are heated at 900°F (480°C) to produce calcium oxide, or quicklime. This is then "slacked" by mixing with water in a hazardous procedure to produce a lime putty that can be added to clay and sand to produce the plaster. Many cultures depend instead on manure, which provides fibers that hold the clay together and enzymes that help plasticize the clay. Fermented and then dried, it has only a pleasant grassy odor.[10]

## The Whole Farm

Such resources have been part of every culture that has lived off the land. American farmers brought some of this sort of knowledge with them and learned more from the Native Americans they encountered. But too often the careful, sustainable exploitation of natural resources gave way to industrial mining of soils, forests, mineral deposits, and bodies of water. And as farming grew committed to commodity production on a large scale, woodlands, woodlots, and wastes were put under the plow or abandoned. Still, for several centuries most American farming relied as much on the whole farm as did traditional cultures.

Given the initial size of American farm properties, which were enormous by European standards at 160 acres for the official "homestead," cropped land was but a small portion of the total. Cropping could grow into forested areas or land still in native prairie grasses, taking advantage of the first fertility of new-plowed land and fertility added by burning off the forest and brush cut to make way for plowing. Meadows and pastures were part of the farm, ensuring low-cost food for working animals and domestic livestock meant for market and table. Some farmers learned to use fire from Native Americans or brought traditions of periodically burning crop stubble, pasture, even woodlands from their homelands. Many kept sheep on wastes and pastures for wool as well as meat, and some sowed flax or hemp for fibers that would be turned into clothing. Pigs were fattened on the masts of fallen hickory and oak nuts in the woods. Orchards and kitchen gardens provided everyday fruit and vegetables and plenty for longer-term storage, including, by the late nineteenth century, canning. But farm families also foraged for edible berries and nuts, cut firewood for heating and cooking, and hunted and fished on their own land and that of neighbors. And most of these practices persisted into the 1950s. In many parts of the country, they persist today.

As forested land gave way to larger and larger croplands, farmers retained sizable woodlots to supply the firewood they still needed. Even on the Illinois and Iowa prairies and farther west, farmers maintained woodlots and orchards as part of the economic strategy of the household. This was the way of life of American farmers, much like traditional farmers throughout the world, with major commitments to crops and livestock central to their practice, but with considerable attention paid to the resources available around them on the larger farm and beyond. New demands and new opportunities shrank the size of the larger farm, at first gradually, then rapidly and drastically in the last half of the twentieth century, as farmers pulled up orchards and cut down the remaining woodlands and woodlots in the race to keep up with uncertain economic opportunity.

Some of the old practices had died out long ago in the face of the new, commercially available conveniences. As early as the 1830s, New England mills turned out inexpensive cloth, first brought to the West in the 1890s through the Sears catalog. At first in the East, then increasingly in even the most remote parts of the country, households gave up locally produced homespun cloth and leather for commercially produced clothing. Many of the tools that farmers and householders depended upon were already manufactured by that time, and eventually metal and plastic would supplant homemade pots and baskets, buckets and ropes. Propane would replace much wood heating by the mid-twentieth century.

Even today with the whole farm drastically reduced as it is, much of the work of the farm household, beyond that directly tied to production, is woman's work—and in many cultures, but particularly our own, it's undervalued, even deprecated and feared. Where today the "successful farmer has a wife with a job in town" (a wife, by the way, who also does the cooking, cleaning, child-rearing, and whatever garden work may be left on the modern farm), traditionally women played an enormously important role in providing the resources of the whole farm for the family's and community's use. *Witch*, *hag*, and *crone* were a few of the terms, rich with opprobrium, for the village herbalist, healer, and midwife. Women in many contemporary farming cultures still fetch the wood, haul the water, tend the garden, manage small livestock, do the cooking, and nurture one and all. They may also be farmers in their own right.

In the West that possibility—that women might be property owners and thus citizens on an equal standing with men—was already erased in the sixteenth century, as the new property laws developed (see chapter 4). At the same time, a campaign emerged on many fronts to limit the public role of women and disparage or even prosecute their traditional work. The witch hunts were one side of this movement. Intellectuals like Francis Bacon added a sinister further twist. Often called a "father of modern science," Bacon might more accurately be called the father of scientism, the unscientific belief that science will lead the human race in an ever-progressing ascent to the utter dominance of nature. To the still-potent objections to mining and draining as a "rape of Mother Nature"—a sentiment reaching back to Roman times—Bacon countered by referring to nature as "a common harlot" who should be "put in constraint, molded, and made as it were new by art and the hand of man" so that humankind could recover the dominion over nature promised it. Her secrets should be wrested from her through forceful "interrogatories" and her inner parts "delved and penetrated" to satisfy our demand for knowledge.[11] For Bacon, agriculture, like all the mechanical arts, was best founded in such efforts to bind and dominate nature, and only effective through literally manly exertions.

Women's arts were submerged in the emerging scientistic world view. But those arts were essential to the household, whether urban or rural. Women have traditionally been the herbalists and basket makers who managed the wild sources for their materials. They plastered the house with mud or manure and sometimes built it themselves of local materials. In traditional American farmsteads women beat the flax, cleaned and carded and spun and wove the wool, made the clothing, and maintained the household. Whether men or women milked the cows, women churned the butter and took eggs and butter to town for extra money, and they preserved food for the winter. In Western cultures and others, women have been variously alewives, bakers, butchers, market vendors, and fishmongers. Some of these roles had prestige, others not. But in much of the work of the larger farm, wherever major emphasis is put on the productive enterprise, today as in the past, woman's work is invisible, except to the extent that she assumes a role in the business of farming. But the business of farming is only part of the whole farm economy that we will have to recover if we are to farm for the long haul.

One example of that ancient economy illustrates the crucial need to take seriously the practices of the past. The recovery of traditional herbal knowledge by both women and men over the last decades promises a return to health care that is sustainable in the long haul. Pharmaceutical antibiotics and their makers are fast losing their ability to keep up with pathogens. Herbal antibiotics and remedies, with their wider spectrum of beneficial effects, don't usually have the spectacular quick turnaround of pharmaceuticals, but they appear to contribute better to the body's inherent ability to heal itself. Herbal medicine was driven underground by the American Medical Association's campaign against all rival approaches starting in the late nineteenth century. The campaign was launched with the explicit promise of increasing the income of doctors by restricting access to certified medical education and carried with it a commitment to pharmaceutical drugs as a chief mode of healing. Hundreds of medical schools were closed, including most that educated women.[12] Today all state boards of medical examiners are in the control of AMA doctors, though chiropractics and osteopathy are still tolerated and acupuncture has gained some respect. Herbal practice was banned in many states and remains so unless carried out by an AMA-approved physician or an acupuncturist. But herbal medicine has made a quiet comeback, as AMA-approved medicine has failed to deliver on its promises and drifted out of reach of most people most of the time.

Today as in the past, many herbs are grown in the garden, but others are wildcrafted. Herbal practitioners are often knowledgeable students of plants, both cultivated and wild. Like their forebears they practice sustainable harvesting and serve as informal monitors of the health of woodlands and forests. In rural communities in particular, they provide everyday healing for millions of people. They cannot rival surgeons in the extreme cases where surgery is called for, but for many people they provide health care that in style and substance is far superior to that of our overpriced, narrowly educated medical professionals. We cannot now afford those professionals, at any rate. In the long haul we will have to rely on our knowledgeable neighbors and the products of our own wilds and wastes for most healing.

Like the women who draw on the resources of the woodlands and wastes for everyday subsistence, the whole farm, the ecosystems beyond cropland and pasture, is generally invisible to productivist eyes. Yet only farms wholly dependent upon external inputs can even claim to succeed without regard to the larger environment, and that claim is specious. The whole farm may be thought of as a series of ecosystems within ecosystems. It is a "farm without borders," as Will Bonsall puts it. Some of the old-fashioned whole farm ecosystem may be brought in by design when we build ponds, construct swales, or plant windbreaks and hedgerows. But the very mentality of "Production first!" tends to render invisible or marginal the wealth of activities that made up the complete livelihood of traditional farmers and much of their subsistence security. In the genuinely resilient farm of the past, today, and in the future, woodland and meadow, kitchen garden and household livestock are all part of the whole farm economy. And large parts of that economy, from woodland soils to field straw, kitchen wastes, and manure, can contribute valuable fertility to the productive enterprise itself at little cost. We'll come back to the economic and social costs of neglecting the whole farm and the whole farm family in chapter 9.

## Fire and Earth

Among the energy challenges that the twenty-first century will bring is the question of how to heat our homes and prepare our meals. Though natural gas is projected to have a longer life span than oil, it will eventually be too expensive to extract. With much of the rural world dependent on propane for cooking and heating, we face a dilemma that few of us are yet prepared to deal with. Few of us can afford the sort of hyper-insulated, airtight, heat-pump-heated home that is all the rage among builders in Europe and theorists in this country. And we live under a political system that will see to it that all the cost of such innovations is borne by us. It is true that in much of the United States, firewood remains an option. But we have long wasted what we use, and widespread recourse to firewood may well threaten our already diminished forests. In effect we are back where our European ancestors were in the sixteenth century when the Little Ice Age, population growth, and urbanization confronted diminished forest resources.

The response then was threefold. Governments and landowners introduced forest conservation measures, restricting firewood gathering and encouraging forest management. Coppicing became more widespread. And energy-efficient tile or masonry stoves were the rage, with builders competing for the most efficient design. Iron stoves, introduced in Europe at the same time, had the advantage of transferring heat rapidly to the room or the cookware. But the stove surface gets very hot, heat dissipates quickly, and so maintaining the heat requires a lot of firewood. Iron stoves, like the fireplaces they replaced, sent most of their heat up the chimney. Though some efficiencies were introduced relatively early, most of us still use designs that suffer the disadvantages of the model. Europeans in the sixteenth and seventeenth centuries built on older masonry stoves, adding baffles and circuitous channels for heat and smoke in order to retain as much heat in the masonry, and thus the house, as possible. These stoves are highly efficient. Where iron box stoves are as little as 20 percent efficient, and the typical

American one is 50 to 55 percent efficient, efficiency ratings for masonry stoves are in the 70 to 90 percent range. Equally important, they burn much less fuel. They are designed to be fired hot and fast once or twice a day, with the heat retained by the masonry and radiated out to the room.[13]

Radiant heating designs are very old. Archaeological evidence from modern Afghanistan suggests that the Afghan style of radiant floor heating could be as much as four thousand years old. We know that Roman hypocausts employing a similar design go back over two thousand years. And Chinese and Korean *kang* and *ondol* date back at least that far. In all these systems, a stone or masonry floor rests above pillars and channels through which the smoke and heat of a cooking stove circulate. If not airtight, of course, the system risks poisoning the inhabitants of the room; but with the proper precautions and upkeep, it provides warmth both day and night as residual heat in the floor lasts a long time. In Russia, Ukraine, and the Baltic region, huge clay stoves inside the main living quarters served a similar purpose. And in many of these societies, the peasant family would sleep on top of the stove in winter weather. Saunas go back much further in Scandinavia, with sweat lodges and related structures common throughout the world. In Scandinavia, however, the sauna stove often doubled as a drying oven and the room as winter living quarters for the family.[14]

The Northern European stoves were designed to be fired with a variety of fuels, from hay to dried dung to faggots. In fact the masonry stove is best fired with smaller-diameter material, lightly packed, which can generate the hot, fast fires the system requires.[15] For this reason the spread of the more efficient versions of masonry stoves in the sixteenth and seventeenth centuries was accompanied by more intensive coppicing of woodlands. A one- or two-year-old coppiced stand could produce abundant faggots for the new stoves, with much less labor than that required to harvest and split rounds of mature trees.

The traditional dual-purpose ovens of Eastern Europe had this commonality with the new masonry stoves developed in Western Europe and Scandinavia: All depended upon the abundant natural resource of clay. Clay ovens date back to the distant origins of cooking, and clay soon came to serve as both a durable container for heat and one that served as human habitation as well. The tile stoves of Europe were only a new variation on this very ancient theme. The older and more rudimentary clay and cob ovens were durable so long as they were protected from the elements and their cracks patched regularly. The masonry stoves developed in early modern Europe depended for their success on careful choice of materials. In particular many builders felt that it was important to ensure that all components, from bricks to mortar to the decorative tile itself, be made with the same clay so that natural expansion and contraction would not crack the stove. The only way to achieve such a standard today, aside from expensive custom work, is to build the stove yourself. That is just what cob cottage builder Ianto Evans advocates. Evans's rocket stove mass heater joins an improvised cooking device with the principles of masonry stove construction to produce energy-efficient thermal and radiant heating for a room. And most of the materials for building it could come from the "farm without walls" available to many people with a bit of land.[16]

Cob ovens have also enjoyed a renaissance. Building on the traditions of Southwest Hispano and French Canadian clay ovens, Kiko Denzer has been building ovens and showing people how to use them for several decades. Like the masonry stoves of Europe, these ovens are energy-conserving, as they start with a quick, hot fire to heat the clay and firebrick floor of the oven. Then the cooking starts: first bread, at up to 800°F (425°C), then pizza or a roast as the oven cools down, then rice and vegetables.[17]

Like Evans, Denzer recommends searching your own landscape first for ingredients. Though the natural building movement has taken a turn in some quarters to much more sophisticated (and expensive) designs and material sources, at its most useful, natural building draws on local resources for durable structures that have the added advantage of being relatively fire-safe.

## Managing the Garden of the Wild

The whole farm depends upon the wild margins that make up the greater ecosystem in which farming has to be embedded. Traditional uses of those wild margins resemble the indigenous practices that have been documented from California to the Amazon to the Australian outback and the rainforests of Africa. Undoubtedly, traditional interaction with the woodlands and wastes that surrounded agricultural communities around the world grew more or less directly out of those practices. Those ancient practices meant that humans, just like any other species, had a distinctive impact upon the world and were integrated into that world. Some of the practices were destructive, but as we have seen in many cultures around the world, peoples who had an interest in providing ongoing resources for human use learned to exploit their surroundings in ways that not only conserved those resources, but at times enhanced them. This is the major lesson of the archaeology of indigenous Amazonia touched upon in chapter 5. It is also a major lesson of Kat Anderson's exhaustive study of California Indian natural resource management.

These lessons fly in the face of a certain prejudice among environmentalists that says that the best way to respect nature is to leave it alone. According to this logic, wilderness will only preserve its value for us if we allow it a value for its own sake, and leave no trace on those rare instances when we interact with it at all. Only by preserving intact areas not touched by human beings can we preserve the biological diversity of our world. But that presupposes that these areas are indeed intact, or will soon recover some version of their pristine condition before human contact if we just leave them alone. The evidence Anderson and others have assembled suggests that this is far from the case for much, if not most, of the natural world. On the contrary, indigenous peoples over millennia tended the wild, enhancing certain features while suppressing others. As indigenous influence was stamped out by colonial governments and settlers, moreover, much of the grace of that natural environment disappeared and much of its balance was lost.

In effect the notion of wilderness reinforces the idea that human beings are alien to nature, inherently destructive, and in need of restraint. Nature, to survive, needs to be confined to gated communities called "wilderness areas" where humans can rarely interact with it and do no harm. But human beings are in fact a part of nature, a product of nature, and have been participants in natural processes from their remotest beginnings. More important, as most farmers know, we can best appreciate nature when we interact with plants, animals, and the soil. As Kat Anderson puts it, reflecting on what she learned from the Native American elders she interviewed, "one gains respect for nature by *using* it judiciously." The elders she talked to maintained, moreover, that many plants do better when gathered respectfully. As any farmer knows, pruning and cultivating can actually enhance the life of the plants we deal with. Finally, she found that many plants and animals had come to depend upon human intervention and management. As indigenous practices were suppressed, many of the most iconic landscapes of California changed for the worse, to the surprise of the Euro-Americans who had taken over.[18]

The lessons of sustainable use that we have seen should inform our thinking about the larger ecosystem upon which our farming depends as much as do the lessons of centuries of despoliation by human users. They should encourage judicious use of the resources that the wild provides and careful management of the ecosystems on which we depend. The growing permaculture literature and practice around food forests is one example of how we might regain the traditional appreciation for sustainable use of woodlands. Likewise, renewed interest in managing woodlands for enhanced wildlife habitat are putting into practice some of the old knowledge.[19]

The sorts of uses of woodlands and wastes that were common even to early American farming families were rarely burdened with the notions of independent family self-sufficiency that are part of our collective imagination about the frontier experience in the United States. That idea, which has powerfully informed even the sustainable agriculture and permaculture movements today, hardly corresponds to the ways in which indigenous peoples, peasants, and traditional American farm families organized their subsistence. In traditional societies there was usually division of labor, sometimes to a considerable degree. One person in a village might be the brick- and stone-mason, charged with supervising the collection of clay and quarrying of stones for building, and skilled at firing brick or dressing stone. Another might be the village potter, another the thatcher. Basketmaking might be shared out among several villagers. Many of the tasks associated with using the available natural resources or turning crops into useful items took group labor. Women would work in teams pounding flax to release the fibers, then spinning the thread to make cloth. Among indigenous peoples, villagers would work together in festive spirit to beat or burn the meadows, flushing out rabbits or quail for hunters to kill. Even the big-game hunt would be organized communally or among a group of hunters.

Similarly, woodlands, wastes, and commons were generally communally managed. Individual families might have claims of long usage on favored gathering spots for herbs, mushrooms, firewood, or basketry materials; hunters might have claims on certain trapping or hunting locations. But the general right of members of the community to forage and harvest the resources of the wild was part of membership in a band or village. And in many cases, the exercise of both family and general rights was strictly regulated by the group, anxious to preserve the commons for everyone. Woodcutting, herb gathering, and mushroom collection might be confined to specified times of the year and strict limits set on household appropriation. Quarrying and mining were often hedged with strict taboos and restrictions. Similarly, European peasants claimed pasture rights on all the village arable lands once harvests were in. In some areas even harvesting grain with a scythe was forbidden; the sickle was enjoined so as to leave as much straw and stubble for general village use as possible.

Such restrictions went with the privilege of access to the greater resources that the community commanded. They were examples of complex systems of communal management that endured, in many cases, for hundreds of years, because they proved sustainable in the deepest sense. And the restrictions that communal management brought were softened by institutions of sharing and festival that enriched everyday life. In the next chapter we consider all the many ways that it takes a village not only to raise a child, but to sustain a viable and enjoyable way of life while managing the resources on which life depends for the long haul.

# Chapter 8

# It Takes a Village: Leisure, Community, and Resilience

Anthropologists are in accord that traditional agriculturalists, pastoral peoples, and hunter-gatherer societies had time on their hands. These so-called primitives, savages, peoples without a State for the most part managed a comfortable living on a few hours a day of work or very sporadic work. Even peasants, forced by lord or market to produce for more than themselves, often had ample leisure time. The anthropological and historical evidence is overwhelming: It was once possible to get a living with a minimum expenditure of time. As anthropologist Marshall Sahlins put it, summarizing the evidence for hunter-gatherer cultures, this was "the original affluent society."[1]

Traditional societies were organized around a principle of sufficiency. Once subsistence was assured, there were no particular incentives to acquire more. Technology was matched to need, and traditional societies were as adept at managing their environment relative to their needs as are we, maybe more so. "The astonishing thing about the Eskimo, or the Australians," says French anthropologist Pierre Clastres, "is precisely the diversity, imagination, and fine quality of their technical activity, the power of invention and efficiency evident in the tools used by those peoples."[2] Adapted tools and well-honed subsistence strategies meant ample leisure time. And leisure time was highly valued. As Clastres writes, even "when the Indians discovered the productive superiority of the white men's axes, they wanted them not in order to produce more in the same amount of time, but to produce as much in a period of time ten times shorter."[3]

Hunter-gatherer societies could get a living in two to four hours of leisurely "work," which they scarcely distinguished from play. But even traditional agriculturalists did not slave in their gardens days on end—not, that is, unless compelled to do so. Rising clan leaders, headmen, chiefs, and kings could call upon farmers to provide for their leaders' households or retinues or for ritual consumption; but too great a strain often provoked rebellion or murder of the offending individual. Even in medieval Europe, where peasants were obliged to produce for the lord's table and trade, there was an astonishing amount of leisure. Sociologist Juliet Schor estimates that between the big vacations surrounding Christmas, Easter, and Midsummer's Eve, numerous saints' days, prescribed rest days, and minor feasts, English peasants occupied a third of the year in holiday. In France and Spain rest may have added up to some five months of the year.[4]

Barbara Ehrenreich tracks the suppression of such widespread human enjoyment through Western history in her exploration of what sociologist Emile Durkheim called "collective effervescence"—ritual and festivity marked by music, dance, and sometimes ecstatic or even riotous episodes. Viewed askance but long tolerated by churchmen, raucous festivity was gradually moved out of churches into the streets by the sixteenth century. About this time, too, the aristocracy and local authorities of Europe began to abandon popular festivities as they adopted new mores of civility appropriate to the touchy life of the royal court. Popular celebrations that featured role reversals and parody of the authorities no longer had the authorities on hand to torment. By the eighteenth century an upstart bourgeoisie had assumed aristocratic attitudes but, even more, insisted on the sober pursuit of profit (for themselves) and work (for the masses) over sociability. "In late-seventeenth-century England, an economist put forth the alarming estimate that each holiday cost the nation fifty thousand pounds, largely in lost labor time."[5] Historians report that "literally thousands of acts of legislation" were introduced over the next three centuries designed to eliminate or curtail carnival and other popular festivities throughout Europe. It was time, European elites felt, for the European populace to take life seriously, an attitude they quickly extended to the natives of the newly colonized lands.

Early capitalists found it hard to compel the incipient working class to accept the discipline they demanded. Max Weber famously argued that a particular strain of Protestantism was enlisted to drive the work ethic required by the demands of capitalism. And it is certainly true that religion played a role in the development of modern attitudes toward work. John Muir's fundamentalist Scots Calvinist father supervised his sons' long days pulling stumps from newly settled Wisconsin forest while reading his Bible. (He may have been a Calvinist but, like Dickens's Gradgrind, he apparently considered supervision the most strenuous form of work to which he was called.)[6] In England the simultaneous enclosure of the commons also had a great deal to do with changing habits of work. Once rural livelihoods were destroyed, desperation made workers of many a peasant. The factory system received the newly landless gladly, and the repeal of England's Poor Laws in 1834 made certain they hired themselves out at whatever rate was offered.[7]

The means for curbing the "laziness" of peasants and natives were various. In eighteenth-century England the new wave of enclosures put thousands off the land. The desperate poor were faced with severe penalties for the slightest crimes. Over two hundred offenses carried the death penalty. It is revealing that many of these were crimes against property, like breaking into buildings to steal or destroy linen or the tools to make it—clearly aimed at the rebellious displaced craftspeople later called Luddites.[8] In the nineteenth and early twentieth centuries, it was common in Central and South America, as in the American South after Reconstruction, to require that Indians, or blacks in the South, prove they were in the employ of some landowner. In Guatemala such vagrancy laws persisted into the 1940s. Precolonial paddy states in Southeast Asia and European colonial authorities around the globe devised endless means to secure the labor of hitherto independent peoples, from outright slavery to the hut tax, which drove Africans under British and French rule to work for white settlers in order to meet their obligations to the colonial authorities.

All these examples suggest just how difficult it was to secure consistent labor from previously free peoples. Our own sense that we just don't have time to do all we need to do stems from multiple complications, not the least of which is the relentless message that we are born for work. French and Portuguese explorers, who rarely applied this message to themselves, were appalled at the "laziness" of the natives of Brazil, who spent large portions of their days lounging or in ceremony. Pierre Clastres comments, "That is what made an unambiguously unfavorable impression on the first European observers of the Indians of Brazil. Great was their disapproval on seeing that these strapping men glowing with health preferred to deck themselves out like women with paint and feathers instead of perspiring away in their gardens. Obviously, these people were deliberately ignorant of the fact that one must earn his daily bread by the sweat of his brow. It wouldn't do, and it didn't last: the Indians were soon put to work, and they died of it."[9] Obviously, an aristocratic version of the later "Protestant ethic" had already extended itself to the privileged adventurers of Catholic France and Portugal as early as the sixteenth century.

## Why We Work So Hard, Earn So Little

Most of us today, Protestant or not, share some such work ethic. Most of us, myself included, insist on being busy, whether it's making ourselves useful around the house or the community or pursuing some meaningful work like writing a book or building up a farming enterprise. And there is certainly a lot that needs doing. But that preoccupation with work can distract from the equally important tasks of "grooming, gossip, and chatter," not to mention festivity, that sustains our social world.[10] Few of us strike a balance that can please both family and boss, friends and fellow workers. But that has as much to do with an economic system that is designed to extract all it can from our work as with the attitudes and mores of late Western civilization.

In the post–Civil War South, the means of extraction were quite crude. Landlords or local merchants (who were often one and the same) were the primary source of consumer goods, seeds, and farm implements. Tenant farmers and freeholders alike were forced to rely upon them for credit at the beginning of the season, and at the end of the season, they generally found that the crop they had mortgaged for credit was not worth their debt. They were bound to the putting-out merchant or landlord for another year, and every year brought deeper debt. When these farmers, white and black, fled indebtedness for the relative frontier of East Texas, they found they faced more distant sources of exploitation. Railroad monopolies and a credit system controlled by eastern banks severely limited the prospects for even cooperatively organized farmers. As we will see in the final chapter, the Populist revolt of the 1880s and '90s grew out of that clash.[11]

With the defeat of Populism, the reforms of Progressivism and the New Deal promised a partial response to the grievances of American farmers; but banking reforms and new farm policy left the basic structures of our economic system in place, and Populism's political defeat left rural America discouraged and divided. Today the structures of our financial system, especially after the banking reforms of the late 1970s, see to it that the wealth created by the many drifts inexorably into the hands of the few. When I started college in 1963, some of my instructors were reading a little pamphlet called *The Triple Revolution*.[12] It was the product of a conference assembled by Robert Maynard Hutchins, former president of the University of Chicago and one of the founders of the great books movement, at his Center for the Study of Democratic Institutions. One of these revolutions was automation and its implications for the economy, work, and the culture. The prediction was that technological advance was raising levels of productivity (product per working hour) so rapidly that the American economy would soon see massive unemployment. What would these workers do? Rising productivity could provide them with a guaranteed income—an idea that the subsequent Nixon administration actually brought to Congress. But how would they occupy their leisure time? The humanists in the crowd were worried; the social activists demanded action.

They needn't have worried. American corporations soon found ways to weaken unions and accrue most of the benefits of rising output per worker to themselves and their shareholders; for the financial sector took the bulk of the wealth created by rising levels of productivity.[13] There followed a famous stagnation in American wages, which still have not budged much above their 1976 levels, and the replacement of the one-wage-earner family by the two-wage-earner family. We have no time, in other words, because we have to work full-time (and increasingly more than full-time) just to maintain the lifestyle of the 1950s—or rather, just to sustain the increasingly fabulous lifestyles of the rich and famous.

The situation has been worse for American farmers, whose incomes relative to the standard of living have been falling since about 1917. One major response has been the parity movement that sought to reestablish the purchasing power of farmers that had prevailed during what some saw as the golden age of American agriculture, from 1909 to 1914. New Deal farm legislation attempted to do just that, and innovations like the Burley Tobacco Growers Cooperative Association, established through the efforts of Wendell Berry's father, John M. Berry Sr., were based on the same aspiration. Charles Walters, late founder of *Acres USA* magazine, was a longtime advocate of parity. The Berry Center (berrycenter.org) continues to promote the legacy of ideas around public policy in favor of parity; but it has little political support in the world that agribusiness and, even more powerfully, big finance have wrought.

One key element of the loss in purchasing power for farmers is the rising cost of production. Relative to the income generated, a late-1940s tractor was a lot cheaper than its 2017 counterpart. The same goes for the petroleum that fuels it, the fertilizers and pesticides, the hand tools, and all those building permits you now need to erect a hoophouse or a cow shed. Add to that odd twist of inflation the exponential growth in what John Michael Greer calls "intermediation," the insertion of nonfarmers between farmers and the dollars spent on food and the food system, and you have a big and rising gap between the farmer's share of the food dollar and everyone else's.[14] From packinghouses to wholesale brokers to supermarkets, from the agricultural extension agents to the university researchers to the bureaucrats who administer "fair marketing agreements" and food safety programs, not to mention the nonprofits that help farmers navigate the rules and regulations and bureaucracies we face—all of them have a claim on the agricultural economy. Factoring in all the intermediaries just to move food from farm to consumer, and including the food processors, the contemporary American farmer gets, on average, just eleven cents on the food dollar.

But another element is farm prices themselves, whose relative decline was mentioned above. US farm "support" policies since the early 1970s have relentlessly promoted overproduction, driving down prices and increasing the burden on taxpayers of paying the difference between production costs of major commodities and market prices. The surplus goes to export, where cheap American commodities have succeeded in wrecking peasant and small farm economies around the world. The policies have kept many farmers in this country afloat, but the depression in commodity prices has made every distortion from cheap processed food to confined animal feeding operations (CAFOs) more profitable than farming. It has also ensured that American farmers "feed the world," that the US market share worldwide stays big enough to subvert the farming sector of countries wherever American commodities are welcome. With such responsibilities, and low and unstable profit margins, it's no wonder we have no time!

## Gaining Time, Growing Community

Leisure is not just an amenity that we can enjoy more or less of depending upon our personal propensity toward compulsive work or self-sacrifice. Leisure is an economic asset. Farming for the long haul requires personal resilience, and resilience means first of all gaining time for ourselves: time to breathe, relax, and reflect. Without such time, farming wears farmers out. Just look at the grim expressions in the old photos! Eliot Coleman remarks somewhere that old-time farmers worked too hard and couldn't sustain a healthy lifetime at it; his goal through a lifetime of farming, teaching, and writing has been to show how it can be done otherwise. Certainly the record of farm abandonment during the heyday of homesteading attests to the hardships of old-fashioned styles of farming in the face of new climates, conditions, and an unforgiving market. And despite the good advice of Coleman and others, the dropout rate for new farmers under thirty in the United States is depressingly high at a time when we need millions of new farmers.

Beyond advice on *how to farm*, we need to take seriously the advice implicit in older cultures about *how to live*. We need to consider the "culture of agriculture," as Wendell Berry has put it.

And a large part of that culture comes from surrounding ourselves with a community that supports our efforts, whether it be fellow farmers and local food advocates who provide fellowship, advice, and concrete support to struggling farmers, or a larger community of eaters who make up an enthusiastic market and source of financial and moral support. And an even larger community is necessary for the sociability, community services and businesses, and collective problem solving that all of us need.

Religious celebrations once helped fill these needs at a time when most people in a given village or region shared roughly the same religious beliefs. Migration and dogmatic battles have destroyed that sort of unanimity and left division and not a little isolation in its wake, inspiring Johnny Cash to sing, "There's something in a Sunday / That makes a body feel alone."[15] Religious festivals, year's end celebrations, and potlatches often served to redistribute wealth within a population, level the social ground between elites and commoners, and build solidarity among members of a community. Civic festivals in our society have never managed to fill that gap, and there aren't enough of them to match the old religious calendars. Spectator sports are scarcely a substitute, and our foodie gatherings aren't inclusive enough to nurture the larger community we need.

If religion no longer serves as a unifying source, perhaps carnival will do. David Fleming, thoughtful economist and Transition Towns mentor, spends a chapter in *Surviving the Future* on the role of carnival—unfettered festivals, ritual disruptions, and plain old merrymaking—in sustaining community. Ehrenreich notes that many observers have seen festivity and group dancing, in particular, "as the great leveler and binder of human communities," and, she goes on, "In the synchronous movement to music or changing voices, the petty rivalries and factional differences that might divide a group could be transmuted into harmless competition over one's prowess as a dancer, or forgotten."[16] And she shows that, despite the repression discussed earlier, collective festivities never really disappeared. Think Mardi Gras. Think rock and reggae festivals. Think Super Bowl parties. Think open mic night at the pub. The more public and the more local, the better for community cohesion. Such events are great equalizers, often subsidized by wealthier members of the community, where community members mix more or less freely. They interrupt routine and give us a break from everyday, sober responsibility. They let us recognize our wilder nature and, in some of their more ritualized or religious forms, Fleming notes, remind us that death and rebirth are part of the natural rhythm of the world. They are indispensable to sane and stable societies unwilling to be consumed by work and worry.[17] We should cultivate them, reward their organizers, support their realization.

Traditional societies also came together around work. The iconic barn raising is the illustration we all know best. Barn raisings are community celebrations, but they are also about reciprocity, the key economic relation in traditional societies. A lot of farmwork even in the United States was once a matter for reciprocal exchanges of labor, usually accompanied by a meal prepared together by the women of the various households. Tobacco farmers gathered to harvest and sort tobacco, sometimes using one farmer's curing house for tobacco from multiple farms. Annual pig slaughters were also sometimes carried out communally, an occasion in some communities for male bonding and a lot of drinking, if Wendell Berry's story "The Regulators" is taken for a guide. And as late as the 1950s, fruit harvesting was often a communal affair, with women, children let off from school for the occasion, and college students on holiday all participating. Before the advent of grain harvesting equipment, such tasks as scything and gathering the grain crop, threshing, and cleaning it were all occasions for group effort, as was bringing in the hay, and farm families went from field to field contributing their labor to their neighbor's crop. Even with mechanization, American farmers in the Midwest and Plains persisted for a time in gang-harvesting their fields, and they still rely on neighbors and friends to bring in the hay in many parts of the country.

Participation in these events, especially in nineteenth- and twentieth-century America, was understood to be voluntary, but nearly everyone volunteered. Not to do so was to signal your social isolation. In many cultures, participation in communal work is mandatory. In traditional rural Spain the term *faena*, and in Latin America *tequio*, describes a communal workday, when members of a community labor together to repair communal infrastructure, build a new *casa comunal*, or provide for those without a breadwinner. As we've seen even today members of Mexican indigenous communities who have migrated to the United States return yearly for tequio in order to retain their membership in the community and thus their rights, from voting or serving in office to access to land. Not surprisingly, the same sort of obligation obtains in the irrigation societies mentioned in the last chapter.

At their core these practices teach the importance of giving and of reciprocity, the essence of traditional economic relations, while guarding against impulses to take advantage of one's neighbors. Western economic theory has obscured these relations, suggesting that traditional peoples met the necessities they could not manage individually through barter, and that when this cumbersome system broke down, they invented money. The market has reigned ever since. Nothing could be further from the truth, as David Graeber shows, summarizing the findings and views of anthropologists, and a few economists, since the nineteenth century. In traditional societies money was just another way of recognizing mutual obligation and relationship, not a medium of exchange for most purposes. And people met their needs through one or another variant on an economy of sharing, often hedged with taboos that forbade equal exchange and underlined the freedom of the gift, and the friendship or mutual obligation that lay behind it.[18]

Marshall Sahlins writes that in so-called primitive societies, material (economic) transactions and social relations are intertwined; the one sustains the other. "If friends make gifts, gifts make friends." And friendship sustains the peace among people who live together or must interact regularly. Thus, treaties are sealed with gifts.[19] Social relations are what govern economic relations in traditional societies, and these can range from pure gift to seizure by force. Among kinfolk or near neighbors, pure gift; among strangers, barter or even open warfare. The model for the first is a mother's care for her children. In most relations among family, friends, and neighbors, even the suggestion of equal return is considered unthinkable because it is insulting to norms of friendship and community. For some societies, traditions of hospitality extend these norms to strangers at the door. Within a larger community, neighbors "lend" to neighbors not expecting return, but anticipating equal generosity in the long run.

This is how many of our community relations work even today, prompting David Fleming to say that this nonmonetary informal economy is "the central core that enables our society to exist."[20] This sort of loose reciprocity ensures that good relations persist indefinitely. Barter, by contrast, traditionally took place between strangers, even enemies, and only works if each perceives an equal exchange in the bargain, despite the fact that each is trying to get the better of the other—a relationship that the money economy and its theorists assume to be the norm everywhere. As Graeber says, "Barter is what you do with those to whom you are *not* bound by ties of hospitality (or kinship, or much of anything else)."[21] Equal exchange may ensure a peaceful outcome, but within one's community one wants more. One wants friendship and generosity.

Stateless societies could be structured wholly on these sorts of reciprocal relations. But centralizing bodies or authorities could also mobilize another sort of reciprocity, what Sahlins calls "reciprocity and pooling with redistribution." In these systems a clan leader, headman, or chief, representing the collectivity, collects and stores goods and redistributes them. Pooling, of course, occurs in the family all the time, but traditional societies often carried it out at the level of the clan or moiety, village, or small kingdom. Virtually everywhere chieftainship entailed the right to a portion of everyone's production and the corresponding obligation of generosity on his or her part. Redistribution ranged from subsidizing festivals and ceremonies to sharing out the spoils of war to supporting crafts production for community use. Among Native Americans of the Pacific Northwest, wealthy individuals held elaborate feasts, or potlatches, during which most of their material wealth was consumed or carried away. The potlatch thus served to redistribute goods periodically among community members and level the social structure, while also reinforcing the prestige of the wealthier members of the community. Saint's day festivals in Mexican villages and barrios served the same functions, draining resources from the rich to the benefit of the whole community.[22]

In settled village societies around the world, reciprocity, not market transactions, characterizes social relations. And nearly everywhere a rough norm prevails that says everyone in the village has a right to subsistence. This usually means subsistence with dignity—that is, having enough to not just get by but to meet communal obligations at festivals and other events, rather than falling into complete dependence. This is what James Scott calls the "moral economy of the peasantry." He attributes it to the precarious existence that supposedly characterized peasant societies; but the sense of mutual aid extends to tribal societies as well, and probably has roots deep in human evolution.

The redistributive mechanisms mentioned above serve to protect the poor from penury and isolation. So do norms that govern the conduct of landlords in peasant societies or tribal officials in indigenous groups, who are expected to take no more in crop or rent in bad years than will permit tenants a decent subsistence, and to practice generosity in their relations with poorer community members. These are not egalitarian utopias; they *are* societies that take care of their own, often because failure to do so leads to backbiting and social strife. The wealthy peasant or landlord who shirks his or her obligations to the community is apt to be the subject of vicious gossip and occasional sabotage. Peasant resistance and rebellion in colonial societies imposing new, capitalist economic relations or onerous schemes of taxation, Scott shows, were, at least initially, about honoring these traditional norms rather than about land reform or radical redistribution.[23]

Once we recognize the true place of reciprocity in economic and social relations, we can begin to appreciate the ways in which we already act it out and recover earlier norms about our obligations to one another. We can appreciate better why we give price breaks to our fellow vendors at farmers market, or share tools and tips with our "competitors," or why our customers will sometimes say, "Keep the change. You guys work so hard." These are the elements of community, and we should cherish their distorting effects on the market. We should also recognize them as economic assets. As Gene Logsdon asks rhetorically throughout his *Letter to a Young Farmer*, "What's that worth?" It's worth a lot, as it turns out. When Hunter and Isa's big home-designed hoophouse blew apart in a terrific storm last year, fellow farmers and neighbors helped them rebuild and a visitor helped them redesign for stability. When Luke and Christilee were looking for land to grow out seeds and conduct variety trials for their new seed company, a local foods group alerted them to the availability of Brookside School Farm. And when several hundred people lost their homes in devastating fires in nearby Potter and Redwood Valleys here in Northern California in October 2017, residents set up resource centers, provided housing and food, and took up collections for temporary relief, daily necessities, and tools to rebuild livelihoods and homes.

In our society women and seniors make up the majority of the caring community. They see to the needs of neighbors, organize the potlucks and fund-raisers for families in distress, and make up the army of volunteers that sustains most of community life. Traditionally, women rarely pursued the hunt for larger game, but frequently helped bring in smaller game and fish. They often were the farmers for the community, as women are in many parts of Africa today. As we have seen, they dressed the skins, beat the flax, spun the wool or cotton, and fashioned most of the clothing and housewares of the community. They nurtured the children and one another, nursed the sick, and served as midwives and herbalists, whether male shamans assumed major healing roles or not. And they generally served as the conscience of the community, sometimes formally, as in Iroquois society, more often informally. In the ancient Middle East, these roles were submerged in the jealous dominance of male heads of households. And in late medieval Europe, as we saw in the last chapter, many of them were demonized as women were pushed out of public life and denied rights to property or even their own persons. But women continue to play pivotal roles as the core of the caring community even in the harshest patriarchal societies, which often sentimentalize them while repressing them.

Once we recognize the true dimensions of our relational economy and the caring community that supports it, and learn fully to value it and reward those who sustain it, we can take steps to rebuild and enlarge it. We can openly attack the presumption, so widespread in American rural culture, that families do not owe a start to their children, that Junior must pay full market price to take over Grandma's place. We can build collective vehicles for providing farmers with access to land, cooks with access to kitchens, and all of us with access to real health care. We can celebrate and enlarge volunteerism and build volunteer buyers' clubs, informal gardening and gleaning co-ops, and information exchanges to enhance access to affordable local foods. We can promote sharing among farmers and neighbors, turn informal tool lending into tool banks, build shared cold storage, exchange seeds and scions, and, of course, find more and more ways to share knowledge.

And more: We can support local food banks, help provide housing for interns and aspiring farmers working rented land, take a hand in securing the welfare of everyone in our community, and support local institutions that provide for the homeless, victims of domestic violence and discrimination, and the elderly. All of this is going on now in many parts of the country and the world, and more will come as we shake off the notion that caring for one another has nothing to do with the struggle to get a living off the land and is best left to governments and nonprofits, or requires credentials, formal institutions, or official status.

## Governing the Commons

Not that communities do not also depend upon institutions. Formal governance provisions are also a part of traditional social orders, essential to making villages, towns, irrigation associations, and communal obligations work. Virtually everywhere, even when chiefs or kings or states claim ultimate jurisdiction, traditional societies have governed themselves in important ways, reserving the most immediate decisions about community life and even the local economy to democratic processes within the community. The New England town meeting is one remnant of that, still vibrant in some places, though stripped of much of its authority by statehouses and their bureaucracies. Alpine villages and traditional irrigation societies still in operation in parts of the world today are self-governing institutions, with sometimes elaborate rules for managing common pastures or water flows.

Governing institutions that work, that have persisted for hundreds, even a thousand years and more, have certain characteristics. For one thing, at the local level they are democratic. That does not mean that their societies are completely egalitarian. Democratic rights may be limited to only some members of the community. It was common in many village societies, for example, to limit voting rights to property holders. These were usually the heirs of founding families in the community, and often males only. Newcomers who acquired property might be admitted to formal membership, and heirs who divided property between them might also be accommodated. Rules varied. But because decisions had mostly to do with the uses of property and the agricultural cycle, heads of households often reserved decision making to themselves. When it came to larger matters, however, like electing town officials or organizing communal work, many village societies extended the vote to a village assembly. In the New England town meeting, for example, all adult members have come to have a vote. Town meeting maintained roads, schools, maybe a library. It elected a town constable and might even maintain a jail.[24] Mexican villages, especially those with an indigenous heritage, often practice this more inclusive democracy.

Governing common resources like Alpine meadows or irrigation systems or fisheries generally requires the more restricted democracy of designated stakeholders only. In fact Elinor Ostrom found that one key to the success of such systems is establishing clear boundaries that define precisely who enjoys rights to the resource. In the case of Alpine meadows, villages own most of the meadows. Rights to graze are restricted to citizens, who are allowed only as many cows on the meadow as they can feed over the winter, and this in turn determines how much cheese each household receives from the cheesemakers. In traditional Japanese villages, decision-making rights regarding the commons are based on cultivation rights, tax obligations, or ownership, varying from most heads of households to only a few. Rights of access are restricted to recognized households. Common property provides villagers with lumber, firewood, thatch for roofing, material for weaving and basket-making, animal fodder, medicinal plants, and so on. Elected officials determine just when harvesting might begin and how much each household can take.[25]

Similarly, in the irrigation systems of eastern Spain, dating back perhaps a thousand years, and in the Ilocos Norte of the Philippines, decision making is confined to those with rights to irrigation water, which are principally determined by property along the irrigation canals. Assemblies of irrigators elect officials annually, determine rules for the distribution of water depending upon availability, and participate in communal work.[26] Similar arrangements characterize the *acequias* of New Mexico, reflecting the Moorish, Spanish, and indigenous heritage of these small-scale waterworks.[27] By contrast, in several of the fisheries Ostrom examined, attempts at governance either did not start or broke down because of the difficulty of maintaining exclusive control of the waters in question.[28]

Other important features characterized successful governance systems. For one thing, states have to acquiesce at least tacitly to the right of users to organize and govern themselves, something lacking in most state-constructed irrigation systems, which consequently have a bad record for serving farmers well. In New Mexico, the New Mexico Acequia Association had to fight for the reversal of a state law that had permitted members to sell water—a privatization of water at variance with acequia principles. Though the state constitution recognized acequia associations as having jurisdiction over their waterworks, the private property bias in US law-making presumed that water was a commodity, eligible for sale by the individual member.[29] Similarly, the inland fisheries of Nova Scotia and Newfoundland, traditionally and successfully governed by local fishers, are endangered by the Canadian state's insistence that it alone has jurisdiction.

Equally important, the commons governance systems Ostrom examined had careful monitoring mechanisms built in to protect them from abuse. In Nova Scotia and Newfoundland, fishers in inland waters have traditionally managed to chase off those without village residence and monitor one another's use of equipment, which has the effect of limiting take.[30] The Alpine villages appoint guards to count the number of cattle as they are delivered to the meadow, and villagers keep an eye on one another to ensure that overwintering rules are honored. Paid monitors figure prominently in traditional Japanese commons management, with guards often exacting a bit of sake for minor infractions. And in the irrigation societies of Spain, the Philippines, and the New World, official water monitors are often responsible for opening and closing gates as each farm comes up for irrigation water. But even when farmers are in charge, both water monitors and neighboring farmers are there to notice premature or otherwise illegal withdrawals.

Finally, successful systems have means of sanctioning those who abuse the resource or the rules, but sanctions are generally minor and graduated, discouraging resentment and encouraging compliance. In Spain a water court hears cases once a week and decides upon sanctions, which start with a few pennies but could lead to expulsion for the rare repeat offender grossly abusing the system. In Japan guards might fine villagers for minor offenses, but more serious offenders could be deprived of their harvest and equipment. They would have to pay a fine to the village to recover the equipment.

Ostrom finds that key to these systems of governance are their democratic institutions, which mean that users have a stake in the rules and they can adapt the rules to their particular situation and to changing circumstances. For that reason, the rules for each of these systems vary from village to village, watershed to watershed, fishery to fishery. And where elected officials have discretion in implementing the rules, as in the Japanese village headman's decision over when to start harvest, users can object by mass disobedience and, eventually, democratic change.

I have spent some time on these systems because they illustrate so well what it takes to manage our common interests. None of these institutions works without organized meetings, decision-making procedures, rules, and considered decisions about the business at hand. They are no more, nor less, complicated than our own contemporary agricultural cooperatives. But many of us shy away from the complexities of organization, the tediousness of meetings, the difficulties of devising decisions that serve us. We mustn't. At a minimum we will need to support the organizations that support us through annual meetings and active attention to the issues that confront us and the people we elect to serve us. Liberty has a price, as Thomas Jefferson famously remarked, and his "eternal vigilance" didn't mean a propensity to grouse about decisions we have left to others.

We will need more, not fewer, self-governing local institutions over the next decades, enriching community among us and providing the sorts of opportunities the state has proven inept at providing. The so-called health care crisis is a case in point. As much a crisis in the sort of health care we are given as in its financing, it is not likely to find a national or state-level solution. But as a community we owe it to one another to forge a response that can serve all our members. Didi Pershouse provocatively notes that, "No one in the United States ever suggests that we should not pitch in to pay for the fire department, because we all understand that any of us could need them at any time. The same is actually true of health-care providers. If your house is burning, you need the fire department. If you are sick, you need care."[31] At least one community has risen to the challenge and created its own self-governing health care alliance, offering care at steep discounts to members, who pay a modest fee to join.[32]

Providing food for all is the natural challenge facing the farmers of the coming decades. With food stamps and the market match program that currently brings food stamp customers to farmers market both on the chopping block, it will be important for more of us to consider how we can give back to the larger community that is our home. Maverick agricultural economist John Ikerd actually suggests we establish "food districts" on the model of school districts and fire districts to pay farmers to provide food for the community.[33] If we manage to move in that direction, the new institutions will have to be a good deal more accountable than their models, but that depends largely on the willingness of ordinary citizens and farmers to get involved. More informal projects, like the FrutaGift free farm stand in Fruitvale, California, or the neighbors' food exchange in suburban Altadena that climate scientist Peter Kalmus describes, can distribute food without the sometimes arcane rules and cash economy of our farmers markets.[34]

We will look more closely at the sorts of economic and political organizations farmers will need to grow as the century unfolds in the next two chapters. For now it's enough to note that democracy is not the hothouse plant of ancient Athens, revived for the world to emulate with the American Revolution. It is the common heritage of humankind, to all appearances our most primitive form of social organization, an impulse deeply embedded in our consciousness. Its exercise is essential to community and essential to providing for ourselves in a world in which the illusion of rational government at the service of all citizens equally has long since lost its conviction.

## Building Community for the Long Haul

Just as much as farmers will need vibrant economic and political organizations, we will need, and need today, the everyday community that includes our friends and neighbors, our buying public, our school district and fire district, our local government, our downtown merchants. Everyday community depends for its vitality on organizations of all sorts, from fraternal orders to food banks. And in the best of cases, these, too, are democratically run. Joining those older community efforts is one thing we can do right now to begin to revitalize and steer our communities toward resilience for the long haul.

Though new organizations exclusively for small farmers, like Greenhorns and the California Farmers Guild, are important for networking, sharing, and building solidarity among farmers, older organizations like the Grange or the Lions Club have roots in the broader community, traditions, and resources that are vital to knitting small farmers once again into the fabric of society. Those organizations once made up much of the warp and woof of our communities. They have been in decline for decades, but many of them are alive and well in rural towns across the country.

Greenhorns founder Severine von Tscharner Fleming called the Grange "a powerful rootstock that we can all graft onto."[35] The sentiment applies to many other community organizations.

Community organizations, like organizations of all sorts, can be difficult to navigate. They are sometimes dominated by local busybodies—most of us who get involved in community affairs tend to have that gene. They can be contentious. They can break apart in ugly feuds. But they are the core of local community. They thrive when small-scale businesses and farms are the dominant economic force in a community.

Community, as historian and social theorist Kirkpatrick Sale has shown, depends on scale. Whether it is the urban neighborhood or the small town, small scale facilitates community. Sale finds "remarkable consistency" among anthropologists, historians, sociologists, and a few maverick economists like Leopold Kohr on the optimal size for human communities. The village or viable neighborhood numbers around five hundred people, enough for what Kohr calls "conviviality," not so many that everyone cannot know everyone else by sight if not by name. A wider group, which might be a tribe or a town, hovers around five thousand people, but in any case below ten thousand. Here, Sale notes, we have a human scale for community sufficient to give people "easy access to public officers, mutual aid among neighbors, and open and trusting social relations. Smallness is simply essential to preserve the values of community as they have been historically observed—intimacy, trust, honesty, mutuality, cooperation, democracy, congeniality."[36]

Other ingredients are important, including small farms themselves. In the late 1940s two congressionally commissioned studies examined the relations between concentrated economic power and the social and economic well-being of communities. Both found that local economies built of lots of small-scale enterprises had a more active civic life, more amenities, and more economic opportunity for more people. Walter Goldschmidt's *Small Business and the Community* contrasted two communities in California's

Central Valley: one dominated by agribusiness-scale farms, the other by many smaller farms. The first had more wage laborers, lower living standards, a more unstable population, and a poorer physical appearance. The second had better schools, parks, and youth services, more religious institutions, and a greater degree of community loyalty. In both studies it appeared that community organizations contributed much to better community life, and those organizations depended for their vitality on small-scale businesspeople and farmers.[37]

Building community resilience often takes place outside the traditional political and community organizations as well. Traveling folk singer and songwriter Dar Williams, a keen observer of the towns she has passed through, has identified some of the elements that help build vital communities. Key ones revolve around food, art, and festivity.[38] Farmers have the food and often a passion for preparing and sharing it, and we often have allies who know how to present it to the public. Some of us have the art, others music. And the festivities go with the food. In other words, if we work at it, have the will, and embrace our allies, we can build community around what we do best. And that will in turn support what we are doing. That is the bottom line.

Community resilience rests on real community, fleshed out in patterns of sociability and vital institutions like schools, voluntary organizations, farmers markets, and communal festivals. And real community can build on and support what we as farmers do best. Living in a rural community to which some of its best and brightest have chosen to return, where others come to find a way of life that the university and the city never offered, is a deep pleasure. Being able to even suspect that one's own work is contributing to the vitality that draws them is an honor. To answer Gene Logsdon's rhetorical question, that's worth a lot.

## Chapter 9

# Getting a Living, Forging a Livelihood

When the blowout of the Deepwater Horizon oil rig destroyed a season's fishing for families along the Gulf Coast, and threatened to impair it forever, President Obama visited the Louisiana coast and reassured the fishermen that his administration would bring jobs to the area. Neither the president nor anyone reporting on the visit seemed to understand the difference between jobs and the livelihoods lost, a way of life that had been central to the lives of most families affected for generations. Indeed, the difference between getting a living and a way of life has been lost on American leaders and the American media for at least the last fifty years.

Politicians of all stripes and both parties (there are still only two, alas!) have been promising jobs while busily destroying livelihoods, sometimes through the same policies. In the countryside, not only have we lost most of our farmers, but we have lost the butchers, millers, bakers, and grocers who provided primary processing and retail services to farmers and small-town dwellers alike. We have lost slaughterhouses, feed stores, hatcheries, cold storage, and even dairies to the seemingly inexorable, but actually planned, growth of national and regional economies. We have seen towns shrivel up and die as farms and the services that they depended upon disappeared, and we have seen the continuous growth of an interstate highway system that was designed to suck wealth and product out of the small places and ended in systematically reducing the product those places could provide the national economy.

To be sure, there are jobs for those who are left, working for the few local businesses that have survived or revived and the many retail giants willing to take people's money anywhere, whether it be the ghetto or a rural town. And in the less devastated areas, people still get by, often by piecing together a living from multiple jobs and tiny enterprises. But there is a profound difference between getting a living and having a livelihood (though we sometimes use the words interchangeably in our careless ways of speaking). The first suggests a scramble, a struggle for survival, dependent on whatever job or sales opportunity turns up. That sense corresponds pretty exactly to the situation many rural people face but also to the situation of the larger workforce. We work to live, and whether work is satisfying or not is a secondary matter.

Having a livelihood, by contrast, suggests a way of life, permanence, and commitment. Living and working are of one piece, and the quality of each informs the other. There may be struggles, to be sure, but livelihoods shape who we are even as we shape how we live and work, and they provide us with periodic satisfaction and long-term identity. Livelihoods spring from community and, in the best of circumstances, vocation or calling.

Traditional farming was a livelihood in this sense, a way of life that transcended market relations and encompassed all aspects of life in the countryside. It coexisted with other, complementary livelihoods that supplied farmers with ropes and harnesses, buckets and hardware, carts and tools, storage and millwork. With the advent of the machine age, more distant makers provided more and more for both household and farm production needs. But with those new goods came a new dependence on the market. Farming became more of a job and the struggle to make ends meet more and more the stuff of farm families' everyday preoccupation. But for many years in this country, farming remained a way of life.

Today the trappings of that older way of life are maintained more as decoration than reality. Rodeos, amateur and professional, feature cowboys who have never worked cattle for a living, and spectators sporting cowboy hats whose work rarely takes them out of doors. Country potlucks feature processed foods of all descriptions and the occasional roasted chicken from Safeway. Country kitsch and country music are everywhere, but real country crafts are mostly the work of hippies, sold at farmers market and crafts fairs. And few families make music together, country or otherwise.

This book has argued that the long haul before us will require a profound reversal of course, from the machine- and petroleum-intense farming that has come to dominate agriculture to a more labor-intense approach; from a farming tied to narrow paradigms of productivity that mine the soil to one grounded in the reality that long-term care for the soil is imperative to our survival; from a preoccupation with production above all to a recognition of the diverse sources of livelihood upon which traditional farming relied. The ground for all those changes can be found in the distinction Wendell Berry and others have repeated frequently, between *the economy* as market or financial venture and *economy* as household management in the broadest sense. We can put the distinction in starker terms, between true economies and false ones.

## True Economies and False

True economies are ones that recognize the long-term costs of everything we do in the name of getting a living. Those costs reach from the personal to the communal, from the health of the land to the sustainability of our extractions from the earth, from our immediate impact to our impact upon the health and safety of the seventh generation. In keeping with the ancient root of the word *economy*, *oikonomia*, true economies are those that contribute to the management of this larger "household." False economies, by contrast, are those that focus narrowly on the bottom line, on production and expanding income first and the kind treatment of earth and others last, if at all. The industrial market economy that dominates our lives is founded on this second, distorted view of human purpose, and its influence affects everything from our community lives to our choice of tools.

*Oikos*, the root of *household*, is also the root of *ecology*, knowledge of the ecosystems of which we are inevitably a part. The false economies of the marketplace not only subordinate those ecosystems to the ends of production, and thus of profit; they ignore them and our impacts upon them as "costs" best "externalized" by the profit-oriented enterprise. The woodlands and wastes of chapter 7 disappear figuratively, if not literally, as we farm for profit at the expense of a genuine livelihood. The pollution that accompanies oil and mineral extraction, plastic and metal fabrication, and transportation and road construction is simply invisible—an unfortunate side effect of growth at any cost, to be remedied (if addressed at all) by a tissue of regulations that scarcely begins to stanch the bleeding of the lifeblood of the planet. A true economy is acutely aware of such costs and strives everywhere to eliminate or mitigate them. As Wendell Berry puts it, we should be reaching for an economy defined by "the making of the human household upon the earth" entailing "the *arts* of adapting kindly the many human households to the earth's many ecosystems and human neighborhoods."[1]

For the last seventy years, if not more, farming has been preoccupied with false economies, and that preoccupation affects the sustainable farming movement as well as what now passes for conventional farming. One of those false economies is the package of practices associated with scaling up. We must scale up, agricultural economists and local food advocates alike argue, if we are to be able to make a living. One prominent adviser, formerly with the agroecology apprenticeship program in Santa Cruz, insists no one can make a living at farming without at least five acres and a tractor. Others shoot higher, and many organic farmers who started out small have seen themselves pressed to grow to twenty, thirty, two hundred acres. There are two strictly market-oriented answers to these voices and one larger point that might be the chief lesson of the agriculture of the past.

The first of those answers is that there are plenty of examples of innovative farmers making a decent living on a couple of acres or less, with minimal to no tractor work. We encountered a number of these micro-farms in chapter 3. Some of them have self-consciously scaled down, adopting more intensive cultivation and a dedication to enhance the fertility of their land. They have done so because there are real economies of scale, but they mostly favor smaller scale. The evidence is unequivocal, though it doesn't permit us to say just what small scale is the most productive, because that depends critically on place, person, and technique.[2]

Second, it is clear that scaling up entails financial costs that advocates frequently neglect to consider. There is that tractor, for example, and the implements it needs, not to mention interest on loans to acquire equipment and the consequent need to deliver returns consistently enough to please the bank. There may be additional land and the challenge of finding it, securing a good long-term lease, and paying for it. The farmer who starts out with just family help or a couple of interns, upon scaling up acquires not just paid employees but also the added costs of employment taxes and insurance. Since scaling up usually means turning to wholesale markets, farmers who started out in mainly retail sales have packaging, storage, and new transportation costs to consider. There is also the considerable waste associated with wholesale production, because those buyers, unlike retail customers, want uniform product. What to do with the rest? And these are just a few reasons why all studies show productivity per acre falls as farms grow in size.

Those are the purely productivist considerations that can make scaling up a false economy. Equally important from the perspective of this book is that it leaves out of account all the contributions to the household economy that the whole farm can provide. In chapter 1 we read Wendell Berry's accounting for his "marginal farm." His account of his income and expenses is worth reading again. It doesn't look like the profit and loss analyses those business advisers are insisting the new farmers learn how to do, but it makes perfect economic sense:

> As income I am counting the value of shelter, subsistence, heating fuel, and money earned by the sale of livestock. As expenses I am counting maintenance, newly purchased equipment, extra livestock feed, newly purchased animals, reclamation work, fencing materials, taxes, and insurance.
>
> If our land had been in better shape when we bought it, our expenses would obviously be much smaller. As it is, once we have completed its restoration, our farm will provide us a home, produce our subsistence, keep us warm in winter, and earn a modest cash income. The significance of this becomes apparent when one considers that most of this land is "unfarmable" by the standards of conventional agriculture, and that most of it was producing nothing at the time we bought it.[3]

Scaling up runs the risk, and too often entails, abandoning that larger farm, what we have called "the whole farm," that provides shelter, subsistence, heating fuel, and much more, not necessarily quantifiable in monetary terms, from great food to good health, in addition to that cash income. It also runs the risk that we lose sight of the "affection" (to use Berry's word) essential to proper care of the land. To the extent it does so, it is a false economy completely aligned with the dominant economic model, no matter that we call our practice "organic" or "sustainable" or "regenerative." Whether and how we scale up will depend upon individual circumstances and goals, but the risks are real, and they have very real implications for the long-term sustainability of our farms. And in the end what we want today to meet the need for local food and a vibrant local economy is not bigger farms but more farms, many more.

These points underline the false economy of a strictly productivist approach to farming. If farming is a vocation, not a mere job, if it offers a livelihood, not merely a paycheck, then it necessarily encompasses more than that paycheck. More important, insofar as farming supports families while caring for the land, it has to go far beyond that paycheck to realize its vocation. It has to embed itself in the local community that supports our work and our family life, as we saw in chapter 8. It has to tend the wild of its own margins and realize the value of the resources the whole farm provides while paying strict attention to the impact of the farming operation on both those resources and the larger ecosystem. And it should support and grow a larger community of farmers and food producers to meet local needs. None of this is on the productivist agenda, another sign that a purely bottom-line approach to farming is captive to the false economy of the marketplace.

It may or may not make sense to hang our income prospects on distant markets. Even if the easy transport of the age of oil disappears, we should recognize that American farmers were supplying European markets well before the advent of the diesel-powered freighter. Nevertheless, the natural context of farming for the long haul is and will be our local communities. Nourishing our community is nourishing ourselves and forms the basis of our livelihood. To quote Berry again, the central question, if we are to have an agriculture for the long haul, is the best way to use the land. And that question does not admit of a universal answer. Instead,

> We are asking what is the best way to farm in each one of the world's numberless places, as defined by topography, soil type, climate, ecology, history, culture, and local need. And we know that the standard cannot be determined only by market demand or productivity or profitability or

> technological capability, or by any other single measure, however important it may be. The agrarian standard, inescapably, is local adaptation, which requires bringing local nature, local people, local economy, and local culture into a practical and enduring harmony.[4]

Anything else is decidedly a "false economy."

## The Tyranny of "the Economy"

We'll come back to false and true economies shortly. But first I want to go back to the question "Why We Work So Hard, Earn So Little." As we saw in the last chapter, American farmers in the 1870s and 1880s had pretty clear answers to that question. In the South, the debt lien system maintained by the furnishing merchants and backed by the banks and cotton buyers ensured that farmers never got enough for their labor to pay for basic provisions, never mind improvements to their operations. In the West and Midwest, the railroads controlled freight rates and thus the profit that might be made on grain shipped east, while East Coast bankers, through tight money policies tied to the gold standard, squeezed both buyers and farmers at harvesttime, relentlessly driving down farm-gate prices.

Our bondage is more subtle. Like the farmers of the late nineteenth century, we face ruinous prices for our products, in our case as a result of a fifty-year-old system of farm subsidies that has produced almost yearly surpluses of commodities, driving down prices to farmers while subsidizing other sectors of the food economy, from processed foods to meat. But it also reflects changes in the cost of living and operating all of us face, from farm families to urban consumers. This is a product of both the growing costs of basic goods, services, and inputs and the growing needs we have accumulated since the nineteenth century. When farmers and their families gave up homespun for manufactured clothing, they acquired new needs for cash and thus for increased farm sales (or that job in town). When they turned from wooden buckets and tubs made by the local barrel maker to purchase tin pails and washing machines manufactured back east, they not only lost the barrel maker and his skills but also increased their need for income. Those transformations have only accelerated since the Second World War, when US policy makers and industrialists self-consciously embarked on a policy of creating a consumer culture. And for farmers they have been multiplied by rapidly rising equipment costs.

Every decade the ante of needs gets upped, and planned obsolescence spins a more rapid cycle. The cell phone is emblematic of the most recent changes. The old product life cycle saw a new product, say a transistor radio, introduced while perhaps still buggy and at considerable cost. Early adopters were willing to pay a lot for the new wonder. Gradually, as manufacturers cashed in on their earnings and honed production, availability increased rapidly while the price went down. The cell phone industry has turned this process on its head. Initially, cell phones were almost free, in exchange for a two-year contract with the phone company. We got hooked, even if the basic phone function remained buggy, all the more so as new advances in electronics delivered more bells and whistles to what was now called a smartphone. Still not smart about phone calls, the new gadgets have become ubiquitous and a virtual necessity for most people, no matter their income, and new models appear even more rapidly than the old phones break down. Sensing the time was ripe, the phone manufacturers and phone companies now charge anywhere from $200 to $800 for a new phone, contract or no. With a new need firmly entrenched, our demand for increased income to satisfy that need has grown exponentially.

That means we have to work harder to earn a living. With neither wages nor farm-gate prices keeping pace with our new needs, surviving becomes a scramble. And for farmers that has meant yet more needs. Land prices have risen and, if we want to keep up, we seemingly either have to scale up, entailing all the new costs outlined above, or buy into the nifty new tools that will allow us to work smarter and produce more in less time.

This is what David Fleming calls a "tight economy." Everyone is under the same pressures, everyone is trying to keep up, innovating where possible, working harder in any case. If you fall behind or slack off, you lose competitiveness, market share, and thus income. The consequence is that the sort of "slack economy" that might allow us to take time off for a sick child, care for our aging mother, or just to participate more actively in community life isn't available to us. Nor do we believe we have time for the kitchen garden, the woodlands, the old crafts of fabrication, preservation, and food preparation that could sustain us in hard times. Building the economy and the community that Fleming sees as our only hope once the current system collapses of its own weight just seems out of the question.[5]

Fortunately, the situation is not quite that dire. The reality is we don't have to buy in. Okay, keep your cell phone. You're addicted and besides, you "need" it. But there are all sorts of ways to disengage, starting with food itself. It is not just a foodie affectation to help your friend slaughter and butcher his hogs, carrying away a small share of meat, bones, and lard. Nor is it romanticism to learn to render that lard and use it for cooking. Along with the meat and the bones for a nourishing bone broth, it's a significant contribution to a living. And if you enjoy the work of harvesting the pork while working with your neighbors, and like preparing your food in the old-fashioned ways, all the better. Many settled agricultural peoples continued to hunt and forage as long as it was possible, regarding these activities not just as supplements to their food supply but inherently pleasurable—more so, perhaps, than farming.

It is not hard to think of more ways in which a subsistence-oriented farm relieves the economic pressure on the household. Backyard chickens or rabbits (provided you forgo some of that expensive feed), household gardens instead of lawns, gleaning those abandoned orchards in the neighborhood, wildcrafting mushrooms and herbs, making use of that woodlot we talked about in chapter 7, and repurposing aging products are all activities that make economic sense, as well as delivering personal satisfaction. As some contemporary Greenhorns put it, "We may live like paupers, but we eat like kings!"[6]

## Tools for Living

We don't have to live like paupers, of course. But curbing our demands for consumer goods and "the American way of life" could have a big impact on what we mean by success in farming. It can also make the difference between pursuing false economies in an often losing effort to keep up and creating a true economy for ourselves and our place. In an essay on "Horse-Drawn Tools and the Doctrine of Labor Saving," published shortly after his 1976 *The Unsettling of America*, Wendell Berry speculates on what American farming and the American countryside, and even urban America, might have looked like had we stuck with the highly developed horse-drawn tools of the 1940s. Not, he says, to recommend the latter over the package of tools and practices that the tractor brought—though he himself has farmed his whole life with animal traction—but to examine what "labor saving means," and, in particular, what "saving labor in the true sense" might mean. In our terms his question was how to distinguish a true economy from the false.

What such a suite of tools might have meant, given the right approach to the vocation of farming, was not the elimination of workers and the diminishing of the skill and care that goes into farming (decidedly a false economy), but a human-scale intensification of production that could have provided leisure to the farmer, as well as time to farm better on all levels. It might also have meant the preservation of livelihoods tied to farming, from carpentry to leatherwork to small-scale metalworking. He ends by insisting that we in fact had (and have) a choice in the technologies we adopt and how we use them. And he insists that "this choice depends absolutely on our willingness to limit our desires as well as the scale and kind of technology we use to satisfy them."[7]

This sort of inquiry, not *necessarily* backward-looking (though this book has insisted that we can learn a great deal from the past), ought to inform our thinking about all the choices we face as farmers. What innovations can allow us to work "smarter, not harder" while enhancing production, our farms as a whole, and our family and community life? Of course, not all decisions have all these implications. But asking such a question can help us avoid false economies and motivate us to search for true ones. And, again, a true economy is not just about production in the narrow sense, but about how our productive activities enhance our staying capacities and our care for the land and its people.

In *You Can Farm*, Joel Salatin recommends tearing down that old barn and throwing up something more versatile, if you need such a structure at all. Gene Logsdon disagrees. His essay on barns in his last book, *Letter to a Young Farmer*, starts with an evocation of the air of calm in an old, active barn. Sitting among the animals in the evening, he experiences "an independence here and a pride that gives a person a kind of satisfaction not common anymore."[8] But characteristically he quickly turns to practical matters. Writing in praise of the possibilities of even the "hobby farm," he remarks, "One never knows when something you decide to raise that seems impractical and unprofitable suddenly shows commercial possibilities. . . . Whoever thought that raw milk would find a viable new market? Or that eggs would again become the darlings of the food faddists, much less butter and cream?" And he notices that "Just down the road from us a roadside marketer and his wife rebuilt the old barn on their property, not only to use as part of their store but to house their workhorses. The barn and the horses both attract customers. How do you figure the profit from that?"[9]

Those old barns turn out to be structures that *Whole Earth Catalog* founder Stuart Brand called "buildings that learn."[10] Logsdon's son-in-law transformed a hog barn into a sheep and horse barn, building mangers and a loft to hold hay. Logsdon built himself a barn on old models, learning how some of the many design tricks developed by older barn builders saved money and made space. He learned, for instance, "that an ear corn crib's narrow, four-foot spacing between walls . . . allow[s] for effective natural air penetration to dry ear corn. The outwardly slanted walls allow moisture to drip down outside the crib not into the corn." Equally important are some of the natural efficiencies built into those traditional structures: "You are hardly ever doing just one thing when you are working in the barn. When you are feeding and bedding down the animals, you are making the fertilizer for next year's crops." Dry bedding makes for happy cows, who reward you with more milk. And the hay you cut and store doesn't have to be bought and provides fertility to the soil from which it comes. "Nothing is wasted. Half-digested corn in the cow manure becomes food for the chickens." The bats in his rafters eliminate mosquitoes and control corn earworm. "What's that worth?"[11]

The versatility of those old barns contrasts with the single-use efficiencies of many of the new market gardener tools. These may be labor saving in the true sense espoused by Berry, but we don't necessarily need the latest and greatest for every farming operation. Eliot Coleman has been a pioneer in devising ways to make farm labor easier for farmers' bodies and more efficient. His colinear hoe does the first, but it's devised for very early weed control only. The same is true of the new tine weeders on the market. Coleman's six-row precision seeder is very versatile as seeders go and one of my favorite tools. My daughter, who runs our salad garden operation, won't use it because our soil is not always suitable for it. And it doesn't easily allow for the sort of intercropping practiced by the Parisian market gardeners. Recently, Ben Hartman has popularized paper-pot transplanters. Characteristically, he is careful to consider when a market gardener might want to adopt such a technology: "If you spend more than a few hours each week transplanting, I recommend giving it a try."[12] You might also want to consider that the paper pots are a continuous expense, and, as he points out, the adhesive in them cannot be used on organically certified farms. All these tools can dramatically shorten the time it takes to do a chore, but they are not for every toolshed. Some can be fabricated at home or by a neighbor—Coleman's books are thorough guides to this sort of hacking. We should emulate him, but in any case we need to ask "Why?" at least three times before we buy when considering a new tool. This sort of questioning needs to go into the real choices we have among the technologies available to us. How we farm is going to be particular to each of our settings, as Berry insists it must be.

The same thinking applies all the more to large-scale technology. It is unlikely indeed that the new GPS-controlled planting systems will outlast the end of the Age of Oil. It *is* clear that they are tremendously expensive, even if they are capable of injecting just the right amount of fertilizer in the soil alongside the seed, inch by inch over hundreds of acres of cropland. They also do nothing to stop the exhaustion of our soils. The new regenerative agriculture has a cheaper, more holistic answer: Attend to the health of your soil, all your soil, through cover cropping, compost, and managed grazing. This, too, may be more difficult on the larger spreads as fossil fuels disappear from everyday use. But the principle is applicable on small-scale farms as well as large, and is as important for vegetable production as for commodity crops. And reconsidering scale will be all-important to returning to the true economy of the best of the older farming. Whether it is crops or livestock, smaller scale permits better care. A closer look at what we can do at the scale at which we operate, like a closer look at the implications of our tools and techniques, is essential to the adaptive farmer in the long haul.

This point extends to overall schemes of production. Those of us who have orchards generally spend a lot of time pruning, and many take special pains to thin fruit or even blossoms to achieve maximum yields. Some swear by dwarf trees or deliberately dwarf them to make it possible to harvest all the fruit worth harvesting. Mark Shepard discards all this, growing widely spaced standards on his orchard-pastures. The fruit that falls as the tree naturally thins itself and the fruit that pickers can't reach eventually go to the pigs, like the chestnuts in those thousand-year-old Corsican chestnut forests.[13] Who's to say that the meat from those pigs won't more than make up for the loss in product from the fruit trees? Perhaps all that labor-intensive, tender love and care is misspent effort, so long as we have livestock to take advantage of the so-called waste. Maybe conventional orchard management is just another false economy. Let's experiment.

## The Future of Farming?

This sort of thinking raises the bigger questions with which this book started. In closing this chapter (and in the next) we go back to them, looking at the options—the very real choices, as Berry would insist—before us, the visions of the future of farming they reflect, and their meaning for farming today. No one knows what the future of farming will look like, of course, because we don't know what the future will bring. That doesn't prevent us from taking prudent provision for tomorrow. And it doesn't prevent us from taking stock of what is on offer today and assessing its worth and likely longevity so we can make our own plans.

A recent issue of retail giant Costco's member newsletter proclaims the "future of farming" to lie in large-scale, high-tech operations, from data-driven corn and soy farms to sterile greenhouse and hydroponic systems. Tremendous amounts of food can be produced, some year-round, with such systems, which are growing more automated by the day. One Nebraska corn and soy farmer profiled manages 1,850 acres with just himself, his father, and one employee. But does all the data he collects (twenty data points per acre) amount to the sort of knowledge of his land that Berry advocates, a knowledge born in affection and guided by care? Can one individual have effective connection with 1,850 acres? When does data, not to mention the bottom line, override affection in such management? When does the rule of science turn our soil into the inert object that data analysis presupposes? Pretty quickly, to judge by the fixation on precise control of chemical fertility that this system espouses.

Equally troubling with these new approaches is the express aim to eliminate human labor from the equation. To be sure, many of the firms involved encourage innovation—among the few employees left. Their efficiencies are not just labor efficiencies; they want to minimize waste of the purchased inputs on which they depend, too, and manage water wisely. Still, the central aim is to reduce the agricultural workforce (what is left of it) to its barest minimum, while producing as much as possible for a world of billions.

But is the problem we face really a shortage of labor? And shall we drive still more people from the livelihoods that give their lives meaning, dignity, and substance? What are those billions going to do with their lives? And do we lack ways to grow enough food to feed our population on the land we have? In an age of robotics and growing unemployment, the answer to the first question would have to be negative. Supposedly, the new technologies will keep us afloat in "food." But will that food reach the millions and billions displaced from still-productive ways of life, if the technological fixes really do churn out the promised cheap food? The reality is that the soil-based versions of future farming are extremely wasteful of what is really in short supply, namely land—and fertile, productive land in particular. Small farms, even micro-farms, as we have seen, can be far more productive per acre than big ones. They can be more profitable. There is good reason to suppose that they can feed the world.[14] And they can be the basis of a culture of care for the land that answers our various hungers, from the purely physical need for calories to the deepest human hungers for rootedness and community.

Not the least troubling aspect of the future of farming so attractive to a business media infatuated with investment opportunities is the dependence of all these approaches on complex systems over which we have none of the traditional control of farming. I have known an aquaponic farmer to lose all his fish, including valuable three-year-old sturgeon, to a power outage coupled with the failure of the double alarm system that was supposed to wake him to the problem. Overdose a hydroponic system with one or another nutrient and the crop is gone. No doubt traditional farming, too, subject as it is to nature's whims, is at risk. Nature can be capricious and her capriciousness devastating. But in the ordinary course of old-fashioned farming, we have ways of coping with natural variability, from diversity in crop choices to agile twists in technique. The automated systems on offer are supposed to be fail-safe. So was nuclear power. When something goes wrong, it can go very wrong indeed. And it will take an army of specialists, earning much more than farmers, to unravel and fix it.

The future of farming envisioned by proponents of a laborless, high-tech future resembles nothing so much as the future fantasies entertained in popular media for the last seventy-five years, from flying cars to houses that clean themselves. Like those fantasies, which have never come to fruition (I remember when our personal vehicles were supposed to be airborne by 1984), many of these systems make good copy for the technophile press but little economic sense. The precision planting systems in use among a few rich (or heavily indebted) farmers add marginal gains to what is already an enterprise with low margins, thanks in no small part to technologically driven and government-subsidized overproduction. The hydroponic and aquaponic companies seem to expand ambitiously and then collapse in bankruptcy, overleveraged in a food system that continues to feature ruinously low food prices. And all depend upon inputs that grow more expensive by the day, including an internet that will be increasingly out of reach as monopolies tighten their grip and the enormous cost of maintaining the server farms escalates.

## Sustainable?

I recently picked up a Natural Resources Conservation Service flyer titled *Conservation for Organic Farmers* and subtitled [*X*] *Farm: An Ecosystem of Sustainability*. It features an eighty-acre vegetable farm with a genuine commitment to ecologically sound practices, from cover cropping, crop rotation, and composting to hedgerows and careful water management. But is it sustainable? No farm that relies on tractors, so far as I can see, is long-haul sustainable, though electric versions, careful recycling, and unforeseeable technological breakthroughs may make such machinery available to the few for a long time.

I have been impressed with the farms like this that I have visited or read about, as I have been impressed with the progress of no-till farming on the huge grain farms of the upper Plains. Their owners are innovative, energetic, and committed to a new way of farming, one that has a real prospect of reversing a long, ruinous history of agricultural destruction of the environment on which it depends. But their farming operations are likely not sustainable. What shall we call them, then, and how should we think about them? I think the best term to describe what they are doing is the one that has become increasingly popular among farmers of this sort: regenerative agriculture. Their practices are devoted not only to maintaining fertility but to regenerating the bases for fertility and biodiversity on their farms. The longer they persist, the more their methods spread, the greater the ultimate benefit for farming and for the planet, because they promise not just to rebuild the capacity of the land for crops but to improve the overall health of the land, sequestering carbon in the process, and encouraging the proliferation of species that make up a healthy ecosystem.

If their practices are not ultimately sustainable in the economic or practical sense, they are laying the groundwork for the small-scale, human-scale farms of the future, which hopefully will continue their regenerative practices but on a scale that is truly sustainable, one dependent upon human and animal power and whatever energy we can collect from the sun itself.

It is possible to envision a different future. There is nothing inevitable about the supposed trends of the technophile alternative. Their failure is more likely than not. But in any case, we can envision a future that is more attuned to the land, more human and more humane, and more—very much more—within the reach of ordinary mortals. As the opponents of corporate globalism have insisted, "A different world is possible."

## Envisioning a Different World

David Fleming points out that as economies grow, so too do all the intermediary services and infrastructure needs on which they depend. Complex bureaucracies, both governmental and corporate;

ports and roads and airports; and specialists of all kinds, from IT to marketing to corporate headhunters. John Michael Greer has something similar in mind when he speaks of "intermediation" coming between us and every product we consume or produce. At some point this intensification becomes top-heavy, weighing down every facet of life, from actual production to a society's capacity for enjoyment. This is the "paradox of intensification," and in Fleming's mind we have passed that point. The coming end of the age of fossil fuels will just exacerbate the downward turn.[15]

In the aftermath, but starting right now, we should be reconstructing the "normal" state of affairs "before the era of the great civic societies, and in the intervals between them," namely "political economies . . . where the terms on which goods and services were exchanged were not based on price. Instead, they were built around a complex culture of arrangements—obligations, loyalties, collaboration—which express the nature and priorities of the community." "No, don't scoff," Fleming adds. "This is what households still do—and friends, neighbours, cricket teams, magistrates, parent-teacher associations, allotment holders." It's the so-called informal economy that in fact holds our society together at the base, secures its human values, and rescues its members (or some of them) from the vagaries of the market economy.[16]

Expanding a genuinely land-based local economy, a true economy, is the challenge, given the demands of the current order on our time and energies, as well as the many ways in which we are embedded in the larger market economy. Somehow we have to endeavor to make a living in this unforgiving larger context while trying to maintain a healthy family life, build community, and construct a different sort of economic order where "it will not be from the impersonal price-calculations of the butcher, the brewer or the baker that we expect our dinner, but from the reciprocal obligations that join a community together, and the benevolence among its members."[17] Too much to ask, perhaps, but this book has laid out some of the elements that can go into our own efforts for the long haul.

First and foremost among these elements is the local economy. We may still take our cues from market prices, but the more we produce and market food and fiber and the products of forest and ocean and streams locally, the more resilience we build into our local community. And that resilience is indispensable for buffering us right now from the storms of the market economy. Wendell Berry makes the same point, with characteristic emphasis on the importance of commitment to place: "An economy genuinely local and neighborly offers to localities a measure of security they cannot derive from a national or a global economy controlled by people who, by principle, have no local commitment."[18] This is also the sort of economy that is most apt to lead to long-term protection of the land and its people.

To ensure that we remain participants in an economy grounded in our communities, it is essential that we survive as farmers, but, as we have seen, that may or may not entail making a profit on the land. There remain lots of ways to make a living, even in our emptied countryside, and therefore lots of ways to sustain a farming way of life so that we can provide for ourselves and make a surplus for our neighbors. Gene Logsdon does not hesitate to embrace what some call hobby farming: "people who garden-farm for the pure pleasure of it while doing something else to make enough money to live on." Some of these hit upon a "hobby" that pays, and they're suddenly "farmers." Some aim to make money from the start, "but intend to keep on doing it, even if they don't." Logsdon sees these folks as at the cutting edge of the local food movement.[19]

Key to staying on the land, I argued earlier, is the ancient principle of subsistence first! Equally important is neighborliness, as Berry calls it, or community as we emphasized in chapter 8. Berry writes,

> So far as I can see, the idea of a local economy rests upon only two principles: neighborhood and subsistence. In

> a viable neighborhood, neighbors ask themselves what they can do or provide for one another, and they find answers that they and their place can afford. . . . Of course, everything needed locally cannot be produced locally. But a viable neighborhood is a community, and a viable community is made up of neighbors who cherish and protect what they have in common. This is the principle of subsistence. A viable community, like a viable farm, protects its own production capacities. It does not import products that it can produce for itself. And it does not export local products until local needs have been met.[20]

In such a community, all the specialized crafts that serve farmers and community members will also flourish, from baking and butchering to carpentry and metalworking. Health care workers, too (provided they aren't addicted to the extravagant salaries of the higher echelons of the medical profession), and teachers and civil servants might all thrive in a healthy local economy.

Economist Leopold Kohr spent a lifetime considering the size of viable human communities. His evidence suggests that for an effective *economic* society with the sorts of basic specializations mentioned above, an optimal size would be a town or quarter of a city of some four to five thousand inhabitants. At this level of population of both producers and consumers, he says, "society seems capable of furnishing its members not only with most of the commodities we associate with a high standard of living, but also of surrounding each person with the margin of leisure without which it could not properly perform its original convivial function [that is of providing people with the companionship that all need]."[21] As we saw in chapter 8, this is also the size that Kirkpatrick Sale finds most conducive to preserving classic values of community.

An urban neighborhood of this size will be necessarily more integrated into a surrounding economy, but it, too, could revolve around urban farms and the small industries that support them and grow out of them. Some of that has begun to happen in the many urban gardening projects in Detroit, and in Will Allen's Growing Power in Milwaukee and Chicago.[22]

Even embedded in the larger economy as they are, many towns are of a size that a local economy can be built and thrive today or in the near future. At their base should be what Berry calls "the land economies" of farming and forestry. Food comes first, of course, and it is no accident that the movement for economic localization has been led in most places by a local food movement. Dying rural towns have revived themselves through art and festival, tourism and retirement options, it is true. But the land economies created most of those towns and must sustain them in the long haul, when the tourist trade may falter and long-distance transportation cannot be taken for granted. And whatever the future, those towns provide the first market for small farmers scraping by in the countryside. The local food revolution has barely scratched that market. As we will argue in the next chapter, it will take organization among both consumers and farmers to build (or rebuild) that market in the face of the current global food system. But that organization is already under way.

And what might the new land economy look like? Two French farmers, Perrine and Charles Hervé-Gruyer, sketch one possibility. Imagine a typical modern organic farm of 100 hectares (247 acres), they write, giving work to a single farmer with a tractor. Put it into the hands of a small team to transform it along something like the following lines, based in permaculture principles and using intensive market gardening as an initial income source. Five hectares might go to preserving biodiversity—a riparian corridor, for example, planted with local species and left without further human interference. Another forty hectares might become an edible forest, planted with dominant nut-bearing trees—walnuts, chestnuts, oaks—in the upper story; woody shrubs like hazelnuts and vines and bushes such as blueberries, currants, raspberries, blackberries in the next story; and yielding as well wild edible plants, from mushrooms to medicinal herbs. The forest, besides food, also provides firewood and lumber, shingles, and materials for fences and handicrafts. The many uses of woodlands and wastes for traditional economies sketched in chapter 7 would become part of the "income" generated by this forest.

Another twenty hectares might be devoted to grain crops in between nut trees—a pasture-orchard of the sort advocated by Mark Shepard. And another twenty hectares will be devoted to milk cows. This, too, will be developed as pasture-orchard, including standard-sized fruit trees. Smaller livestock, foraging in the edible forest, the pasture-orchard, and the pastureland, will provide meat. Finally, the remaining fifteen hectares could be divided among as many market gardeners, providing abundant vegetables and ready income while the rest matures.

The resulting transformation would employ up to thirty people in agricultural work. It would also generate the need for numerous craftspeople "building and repairing the tools of the farm . . . building and repairing the lodging and infrastructure . . . work[ing] with materials generated by resources in the agroecosystem: weaver, potter, wood turner, joiner, carpenter." The Hervé-Gruyers suggest still more elements—from animal traction to a biogas plant, food processing, restaurants, and agritourism—but the outline is of more interest than the details. With the right energy a modern farm, conventional or organic, could be transformed into an oasis of abundance—and livelihoods—for two or three dozen farmer-artisans and their surrounding community.[23]

Is this vision far-fetched? Or is it the farming of the future, farming for the long haul, that so many of our efforts and longings point to? It offers, whatever we think of the details, the measure of hope that so much else in our current situation seeks to deny us. Its prospects depend only on our will to make it happen and nurture it to success. That may take a village, or a movement like some of those we explore in the final chapter. But it's not outside our reach. And it partakes of the conditions of a true economy, one rooted in local production for local needs, "the making of the human household upon the earth . . . adapting kindly the many human households to the earth's many ecosystems and human neighborhoods."

Whatever its specific form, the small farm economy we need to be fostering will build on the long heritage of indigenous, peasant, and traditional livelihoods, their methods, and their ways of life. Ultimately, the fossil-fuel-dependent industrial model is just a short blip in the long arc of human dwelling on the land, though one with far-reaching and disastrous consequences for the land itself. The knowledge and lifeways of indigenous, peasant, and traditional farming cultures provide a sturdier foundation for our efforts to build a sustainable and human economy now and for the long haul. Those lifeways were not invulnerable nor infallible. They have always been challenged by natural disaster, political domination, and economic upheaval. But as we will see in the final chapter, they have found ways to respond that give them a resilience and a staying power that our present, all-too-brittle civilization lacks.

## Chapter 10

# Farmer, Citizen, Survivor: Politics and Resilience

*We've already had situations where we've waited for the central government to come and help us. And we've waited in vain. We've waited and waited. So, this time, we did not wait. What we did was unite the whole community. We know what we need. We know what each of us requires. So we established five committees. . . . We cannot wait for someone to come from outside, because of the urgency. They've left us a week without anything. There's no supplies, nothing. We need to take care of ourselves. And that's what you see here.*

—Survivor of Hurricane Maria, Puerto Rico, on *Democracy Now!*, October 2, 2017

The hurricanes and fires that left swaths of destruction in Texas, Florida, the Caribbean, and Northern California in late 2017 underline the precariousness of even the most adaptive and resilient of communities. There is little we can do in the face of disaster except cope. Many coped well in these emergencies, and some were prepared better than others. But disaster can overturn one's carefully nurtured livelihood in a matter of minutes. At that point personal resourcefulness and community responsiveness are the best that human resilience can provide, because the governments we thought we could rely on are usually too clumsy to make an immediate difference. Long before that, as these disasters also showed, public policy decisions often make the difference between catastrophic loss of life and livelihoods and relative resilience. Rampant development in Houston's floodplain, the centralization of Puerto Rico's rickety electrical grid, a century of fire suppression in the West—these were public policies that contributed to the immense scope of the floods, isolation, and fires that swept these areas.

American farm culture is built on a myth of self-reliance, and indeed self-reliance can be a powerful tool in everyday life and a resource in the face of disaster. But self-reliance *is* a myth: None of us can survive without the larger community. When government failed to respond, or respond adequately, as catastrophe unfolded in the Caribbean, Houston, and Northern California, individual and volunteer efforts were vital and continued to be important months afterward. All of us face forces beyond our individual control, and each of us stands to benefit from organizing for common ends. Whether the aim is disaster relief, economic betterment, or addressing a political system otherwise deaf to our concerns, we need one another now and for the long haul.

In this chapter we look at the varied ways farming peoples have dealt with the economic and political forces that operate beyond their immediate control and shape even the natural forces with which they have to contend. Economist Albert Hirschman famously described two ways in which people might deal with situations they dislike. One is "exit," the other "voice." Each one is a distinctive alternative to "loyalty." But loyalty should not be confused with acquiescence, the resigned conformity of workers or citizens who know they have little other choice. Under these circumstances, we might find another sort of response, perhaps a subset of "voice," but a subversive one, what political scientist James Scott calls "weapons of the weak." We'll look at each of these options in turn.

## Exodus as Exit

The story of the escape of the Jews from Egypt is one of the best-remembered remnants of what was once the common culture of the West. The details of who these people were and what they became are perhaps less known to most people today than the drama of their flight. The biblical account tells that they were herders who settled in Egypt. There they were eventually reduced to bondage, working as day laborers in the fields and the projects of the pharaoh. Their dramatic escape eventually led them to what is today Palestine, where they occupied the hill country, becoming farmers and settled shepherds. The independent kingdoms that grew out of their tribes flourished, then foundered, until much of the population was carried off to Mesopotamia by the Babylonians under Nebuchadnezzar in the sixth century BCE. In the seventy years of exile, those that remained forged a new life, and when the exiles were allowed to return by the decree of Cyrus II, they came back to resume the old life on new terms, their sovereignty gone.

The stories of exile and return that occupy great parts of the Hebrew scriptures are stories that characterize many peoples historically. Exploited for their product and their labor, peasants have often fled, moving back and forth, in some cases between nomadism and settled agriculture. Patterns of exodus accompanied the decline of many agricultural civilizations, as farmers simply drifted off in the face of drought and growing civil strife. Ancient Pueblan civilization devolved into the scattered pueblos and river valley settlements that the Spanish encountered on exploring our Southwest. Mayans left the warring city-states of the Yucatan and

Petén Peninsulas and settled in the hill country to the west, in what is now Chiapas and Guatemala. And most of the hill tribes of Southeast Asia, James Scott shows, were refugees from the tyrannical paddy states of the lowlands or from the spread of Han settlers and authority in the Chinese empire.[1]

Exodus often brought new forms of livelihood and new identities. Pastoral nomadism from Mongolia to the Middle East was probably not a primitive outgrowth of hunting and gathering cultures but a convenient way for groups to evade the authority of emerging agricultural states.[2] Nomadic tribes developed a distinctive way of life around herding, but they also often engaged in raiding more settled societies, and as some came to specialize in warfare they periodically conquered agrarian societies and settled there, becoming a new aristocracy over a subject population. So the Normans conquered England and established the feudal society we associate with medieval England. So the Assyrians overran Mesopotamia, one of several waves of invaders. But other escapees from captivity in the agrarian states of the past did not develop the option of military dominance. They became the hill tribes who learned to farm on distinctive terms, returning to swidden agriculture even while adopting new crops and methods. In Southeast Asia, at least, there is ample evidence that ethnicity and language alike developed along their own lines among the refugees who gathered in the safety of the mountains and forests.[3]

Exodus resonates in our culture, even today, because much of the settlement of the United States was experienced as an exodus from tyranny, precarious living conditions, or overcrowding. Oscar Handlin's classic study of European immigrants to the United States draws in broad strokes the situation of peasants in an overcrowded Europe; and the portrait applies to the circumstances of many immigrants.[4] The impulse to simply move on in the face of limited opportunities at home fueled the westward migration of both these and earlier settlers and informed our own culture of mobility.

Exodus may be an alternative to captivity, but it is also exile. And exiles settle uneasily on the land and often find their former experiences less than helpful with new soil, a new climate, new conditions of production, and new markets. They leave behind their long experience of stewardship, if they enjoyed it at all, and they are too apt to move on again rather than cultivate the soil and the society where they find themselves. They can lend diversity and richness to the places they come to, but it takes years, even generations, to grow the sorts of roots that are required to tend the land well.

As Wendell Berry says, genuine stewardship lies "in the possibility of settled families and local communities, in which the knowledge of proper means and methods, proper moderations and restraints, can be handed down, and so accumulate in place and stay alive; the experience of one generation is not adequate to inform and control its actions."[5] Thanks to the relentless uprooting that a national economy and an education system focused on upward mobility has thrust upon us, and to our own immigrant roots, most Americans are exiles, and those of us who choose to recover the sounder principles of caring for land and community are only slowly learning to be rooted. We should avoid exodus where we can. We will need a culture that rewards and encourages rootedness instead of mobility if we are to assume a role as proper stewards of the land and truly farm for the long haul. But that means that we will need also to cultivate voice as our first and most persistent response to the larger forces that attempt to shape our destiny.

## Stifled Voices and the Weapons of the Weak

Preoccupied as we are with the art of nourishing life on our little plot of land, most of us find little time to deal with the policies and programs that shape our own possibilities, not to mention the character of the larger world. But the direct effects of decisions made elsewhere on our lives as farmers and food producers are felt every day. The capacity of the state to care for its citizens may be waning, but its hunger to regulate and police seems unabated. And beyond the annoyances, and sometimes threats, of regulation, we face a monetary system stacked against us; a financial system on the brink of disaster and with little to offer struggling small farmers; a commercial system ruled by behemoth corporations; and a food system that puts every farm and farmer at risk of failure. In the background, and sometimes right in front of our noses, climate change threatens to undermine all our efforts to thrive on the land.

Many of us feel helpless in the face of all these threats, and I can't claim that there is much we can do about them. Voice, the effort to speak out and have a say in affairs that trouble us, depends upon a sense that we have a chance of changing things. And that depends, in part, on a sense that others are joined with us in the effort, that collectively *we* can make a difference. Our political system and much of our upbringing seem to conspire to deprive us of those expectations. If social movements in the recent past have brought about changes for the better in our society, we live at a moment when most people, most of the time, feel there is little we can do to bring about significant change. We are better off than many farmers of the past, caught in systems of oppression that offered still less hope for release. But many of our responses mirror theirs.

In a Malay village, James Scott found beneath the surface of peace and civility a peasantry seething with resentment at the landowners and merchants who were the chief beneficiaries of the villagers' labors. And that resentment took form not in voice, nor in exit, but in what Scott dubbed "weapons of the weak"—veiled resistance and subversion meant as much to vent frustration as to actually change the situation peasants faced. "Here I have in mind the ordinary weapons of relatively powerless groups: foot dragging, dissimulation, false compliance, pilfering, feigned ignorance, slander, arson, sabotage, and so forth." Such forms of resistance "have certain features in common. They require little or no coordination or planning; they often represent a form of individual self-help; and they typically avoid any direct symbolic confrontation with authority or with elite norms."[6]

Subtle resistance expressed through such weapons of the weak is widespread in American society, as in most of the thoroughly governed societies of the contemporary world. The immense disrespect in which authorities are held, expressed in gossip and backbiting and, above all, grousing, is a sign that our promised government "of, by, and for the people" looks like anything but to many people. Equally troubling for a polity supposedly founded upon the rule of law is widespread defiance of the regulatory state. Though common enough in urban settings, rural areas seem to be especially rife with otherwise honest folks thumbing their noses at the regulators. Tradespeople working off the books, contractors and homeowners forgoing the building permit regime, farmers diverting water and building ponds without required permissions, and poaching of all sorts are simply a small part of the everyday acts of resistance to rules that seem onerous and systems of enforcement that just don't work. No one is particularly embarrassed by it. And none of the bureaucrats or legislators so ignored seem worried about the effects that unenforceable laws have on the public respect for law that is supposed to undergird our republic.

Such weapons of the weak—like their inverse, resigned compliance—do little to change the established order. They do articulate a culture of rejection that might one day be turned to more positive action. Until that day comes, however, they are simply a way of coping with a system of rules that seem designed more to stifle initiative and enterprise than to promote well-being. Liberals and progressives have no conception of how to respond to large-scale public rejection of much of what government actually does, preferring to ignore it rather than bow to the "conservative" notion that big government might be, well, too big after all. Conservative politicians love to play on our discontents while throwing their support behind deregulation for the corporations that run amok even under the thumb of the regulatory state. Most people, politicians included, have no good idea what makes bureaucracies—both governmental and corporate—so onerous. In the meantime most of us just cope and struggle to explain how citizens of a supposed democracy can justify their disdain for the rule of law.

The regulatory state, moreover, is just one part of the oppression we face—because the government conceived as "for the people, by the people, and of the people" long ago found its central purpose in fostering a national economy that provided the basis for corporate power. From the tariff schemes of Alexander Hamilton through the monetary disputes of the 1830s and 1880s, from the drive to join the continent with railroad and telegraph to its modernized counterparts of internet and an interstate highway system, federal and state governments have been devoted to enabling the growth of a system of national commerce that has swallowed the local economies of the settlers and destroyed the farming economy on which our small towns and cities were built. As usual, Wendell Berry lays out the choices before us with profound simplicity. If government will not protect us from an economy dictated by the large corporations, he writes, then we have recourse to "a venerable principle: powers not exercised by government return to the people. If the government does not propose to protect the lives, the livelihoods, and the freedoms of its people, then the people must think about protecting themselves."[7] Protecting ourselves from misplaced government as well as from overweening corporations requires more than foot dragging and subversion. As Berry writes, it means building local institutions of the sort discussed in the last two chapters. It also means recovering our voices.

## The Farmers' Voice

Farmers have apparently always been a quiet lot. The veil of consent that Scott found in his Malay village was common enough to persuade even that redoubtable revolutionary Karl Marx that farmers are useless for purposes of revolution, so many "potatoes in a sack," inert and impossible to organize (some of us have had that experience!).[8] But from time to time hardworking farmers have risen up to try to change the systems that make us work so hard, earn so little. And in doing so, they have taken a stand for enduring human values of fairness and small-d democracy; they have also shaken the foundations of their societies and sometimes profoundly reshaped them.

Wherever the exit option was closed and the livelihoods of traditional cultures threatened, indigenous people, peasants, and farmers have turned to revolt, both peaceful and violent. In peasant Europe these revolts often grew out of rowdy popular festivities. And their claims were equally often radically egalitarian. The German Peasants' War of 1524 came as landlords were demanding labor dues from free peasants to satisfy growing demand from the cities, and it started in a carnival mood. Martin Luther, whose sympathies the peasants had counted upon, supported instead the princes and aristocrats on whom he depended for his defense against Rome. Thomas Müntzer's response to Luther and the princes was to raise a peasant army against all authority and all property beyond that needed to provide home and food for ordinary people.[9] The English Peasants' Revolt, or Great Rising of 1381, came in response to an onerous flat poll tax imposed on a countryside reeling in the aftermath of the Black Death. When Wat Tyler led peasants to London, destroying tax records as they went, the priest John Ball famously asked: "When Adam delved and Eve span, Who was then the gentleman?" and explained, "From the beginning all men were created alike, and our bondage or servitude came in by the unjust oppression of naughty men."[10] For his sentiments Ball was hanged, drawn, and quartered that same year; Tyler was wounded in a struggle with King Richard II himself, and eventually killed in his hospital bed, but not before rebel peasants dispersed homeward with promises of redress of grievances.

The Diggers, led by Gerrard Winstanley, set out in 1649 to occupy a barren hillside, farm it collectively, and compel the leaders of the English Revolution to repudiate the evils of private property. "The earth was not made purposely for you, to be Lords of it, and we to be your Slaves, Servants, and Beggers, but it was made to be a common Livelihood to all, with respect of persons," wrote Winstanley. The rebellion against private property was late in coming. Already by then more than two hundred thousand gentry, yeomen, and tenant farmers claimed enough property to be able to vote; they made up the bulk of the Parliamentary Army that unseated King Charles I and eventually saw him beheaded. And among them the Puritan Levellers had begun to promote the radical idea of human equality. "To every individual in nature is given an individual property [in himself] by nature, not to be invaded or usurped by any," as Richard Overton wrote. The Levellers advocated near universal male suffrage on this basis, but they also embraced private property.[11]

Our own Whiskey Rebellion was the revolt of farmers in western Pennsylvania whose only cash crop, corn liquor, was suddenly threatened with a tax created to cover the nation's war debt. Protesters stormed the fortified home of a chief tax collector in July 1794, prompting Secretary of the Treasury Alexander Hamilton to persuade George Washington to raise an army against them. President Washington himself rode at the head of the force, which was larger than any army he had commanded during the Revolutionary War. But the protesters had dispersed before he arrived. Many were war veterans, defending the principle of "no taxation without representation," with Hamilton and Washington defending the federal government's prerogative to raise taxes as Congress pleased.

The possibility of success for these sorts of spontaneous revolts probably closed two or three hundred years ago in the more developed states, and as we saw most were quickly disposed of by the weak states of the day. Displays of power like that mobilized by

Hamilton and Washington could quickly lead to the conclusion that there was not much ordinary folks could do to take matters into their own hands. Protest has persisted into the present, with tractor marches on Washington in the 1980s, but sociologist Charles Tilly has shown that advanced democratic states, every bit as much as totalitarian ones, have high capacity for suppressing violent protest.[12] Protests today are mild affairs, parties on the Washington Mall on a Saturday or Sunday afternoon more than an effective display of outrage. Our marchers rarely even venture to try to paralyze the system by filling the jails, as the civil rights protesters did in what was perhaps the last truly effective national protest movement in our history. But over a decade of "endless war" has drained the treasury, the financial system remains on the same shaky ground it occupied before the collapse of 2008, and our political system is nearly paralyzed. Ferocious as it may be, even the wealthiest national state in the history of the world is not invulnerable.

Larger movements for change, rooted in the discontent of farmers, have shaped and reshaped our world. From the nineteenth century on in weaker states around the world, and even in this country at one time, farmers began to organize in movements for large-scale structural change. In Mexico a revolt by liberal elites against the dictatorship of Porfirio Díaz in 1912 turned into a full-blown revolution and civil war when Emiliano Zapata led peasants in a revolt for land and Pancho Villa organized northern cowboys and railroad workers around egalitarian claims. Zapata's rebels came from the state of Morelos, where massive sugar plantations had repeatedly encroached on traditional village lands, and landless peasants faced miserable working conditions and wages in the sugar refineries. With the slogan "land to those who work it with their hands" (*la tierra volverá a quienes la trabajan con sus manos*—the official slogan of the state of Morelos, where the ruins of many of those refineries are still visible), Zapata's peasants fought first the forces of Díaz, then those of the liberal Constitutionalist government until Zapata himself was assassinated and the government conceded major land reform in Morelos and neighboring states. Commitment to land reform transformed the face of Mexico over the first half of the twentieth century, though agricultural policies eventually reduced peasant livelihoods to poverty levels through much of the country.

The Mexican Revolution was just the first of twentieth-century civil struggles in which peasants played powerful roles, from Russia and China to Vietnam, Nicaragua, and El Salvador. Though many of these rebellions overthrew onerous governments, as in the case of the Mexican Revolution few genuinely raised the welfare of farmers, and some, like the Russian and Chinese Revolutions, ultimately subordinated peasants in still more onerous dictatorships.[13] Scott echoes most of the voices of the sociologists and historians who chronicled and analyzed these movements when he notes that peasant victories are few and far between. There may be gains, concessions, a brief respite, "and, not least, a memory of resistance and courage that may lie in wait for the future. Such gains, however, are uncertain, while the carnage, the repression, and the demoralization of defeat are all too certain and real." Even where successful, the results are "at the very best, a mixed blessing for the peasantry. Whatever else the revolution may achieve, it almost always creates a more coercive and hegemonic state apparatus—one that is often able to batten itself on the rural population like no other before it. All too frequently the peasantry finds itself in the ironic position of having helped to power a ruling group whose plans for industrialization, taxation, and collectivization are very much at odds with the goals for which peasants had imagined they were fighting."[14] This is a pessimistic judgment, and it embodies the modernist prejudice that the age of the peasant, and of small-scale agriculture generally, is over; but history has not borne out either the judgment or the prediction, as we will see.

## The Struggle for Economic Power

Farmers' voices have not always led to bloodshed or aimed at complete overturn of the existing order. Our own agricultural history provides plenty of examples of large-scale efforts to defend farm livelihoods through collective action. The Grange (or Patrons of Husbandry) was founded after the Civil War in large part to answer the need of midwestern and southern farmers for help with marketing, equipment purchase, and technical assistance. The Grange was instrumental in bringing the cooperative movement to American farming. But the Grange also fought a series of economic and legislative battles against the monopoly power that enabled railroads to charge what they wanted to ship farm products to market.

Under the leadership of its founder, Oliver Kelley, the Grange fought these battles on two fronts. First, Kelley instituted the figure of the Grange Agent to negotiate discounted rates on farm equipment for Grangers and market Grange farmers' products. In that way the Grange broke some of the power of the railroads over costs and prices. This possibility, and the rapid spread of cooperative Grange stores, grain elevators, warehouses, pork packinghouses, and flour mills, spurred a rapid growth of Granges in the Midwest in the early 1870s. But the second prong was equally attractive to farmers. Kelley and his allies encouraged Grangers to participate in politics and to advocate for regulation of the railroads. In 1871, with major support from a strong Minnesota State Grange, Republican governor Horace Austin won passage of a law empowering a railroad commissioner to investigate abuses and a second law that established maximum rates for freight. Minnesota's laws were followed by similar ones in Iowa, Illinois, and Wisconsin, where the Grange also played a large role in passing them. The so-called Granger laws, challenged in the Supreme Court and found constitutional, set the pattern for similar laws across the country.

Unfortunately, Kelley never had the support of the National Grange officers, men he had handpicked as fellow founders of the new organization. Their leader, William Saunders, a Department of Agriculture official, actively opposed cooperatives and political activism. When the National Grange passed a rule that "No Grange, if true to its obligations, can discuss political or religious questions, nor call political conventions, nor nominate candidates, nor even discuss their merits in its meetings," the radicals and the many farmers they represented were confronted with confusion and division at the local level. That and the failure of some of the Grange stores and cooperatives, charges of corruption leveled at Grange business agents, and partisan infighting brought the spectacular growth of the Grange to a halt, and the organization began its long decline.[15] Today the Grange may openly advocate for policy change on a nonpartisan basis, but the National Grange continues to be a conservative brake on more innovative efforts at the local level. And in many Grange Halls, older Grangers resist the influx of newcomers that the new farming movement has brought.

The conditions that Oliver Kelley and his fellow radicals addressed did not disappear with the Granger laws. Railroads continued to exercise monopoly power in many parts of the country, and merchants often had the same sort of power to set prices for farm equipment and necessities at the local level. In the South, as we have seen, the so-called furnishing merchants advanced farmers credit for necessities with a claim on the harvest. Come harvesttime the merchant sold the crop and regularly reported that the income was not enough to cover the debt, which would have to be met with next year's crop. The resulting debt peonage left tens of thousands of white and black farmers, owners and tenants alike, in poverty. Equally troubling, the deflation that the national economy had experienced since the Civil War, while benefiting the holders of government bonds, exacerbated the difficulty farmers had making a living, systematically reducing farm-gate prices year after year over a long thirty years.

Farmers in East Texas responded in the 1880s with the creation of the Farmers' Alliance. Like the Grange, the Alliance gained momentum through schemes for cooperative buying and selling. The Alliance co-ops, unlike those of the Grange, extended credit and thus proved much more attractive to hard-strapped farmers in the clutches of the furnishing merchants. But neither bankers, wholesalers, nor buyers were willing to extend credit to the Alliance stores and warehouses. As the cooperative scheme foundered, Alliance leaders began to see the plight of farmers in broader, more political terms. As in the Grange, part of the leadership opposed the political turn, but all understood the political and economic challenges, and in this case the radicals won the day.

The Farmers' Alliance, and the Populist Party that grew out of it, represents the most far-reaching and tragic democratic insurgency in the nation's history. Like the Grange before it, the Alliance grew rapidly as it pursued both cooperative development and the new political agenda. After the Alliance promulgated a radical platform in 1886, it spread rapidly throughout the South and into the West. Political radicalism grew naturally out of the experience of trying to establish the cooperative system that the Alliance had pioneered in Texas. In the South cooperatives and state-level exchanges found themselves opposed by local merchants and bankers, wholesalers and cotton buyers, long used to extracting every penny from their commerce with indebted farmers. In the West railroads, grain elevator companies, land and livestock companies, and bankers mounted open opposition to the new cooperative enterprises. In the West, too, recent efforts to create third parties around the greenback platform (aimed at ending deflation and liberalizing the monetary system) found the Alliance an exciting organizing tool. By the late 1880s the National Farmers' Alliance and Cooperative Union had reached all across the South, with alliances in most counties, and into Kansas, Oklahoma, Nebraska, Iowa, Illinois, and beyond.

But the Populist movement launched by the Farmers' Alliance faced division within its own ranks, sectional loyalties to the existing parties bred of the Civil War, divisions between whites and blacks in the South, and an urban society not convinced by the farmers' greenback radicalism. Trying to forge an alliance among all the "laboring classes," the "plain folk," the Alliance was crippled by the lack of effective working-class organizations and the incomprehension of Catholic immigrant workers before a movement led by white Protestant farmers. National president Charles Macune actively opposed formation of a third party, but put all his hopes on his own scheme to generate both credit and currency by establishing government-owned warehouses where crops could be stored until sale in exchange for greenbacks. Neither of the existing parties would take up the challenge until well into the twentieth century, and then only partially and only for the better-off farmers among those who had survived the great economic purge of the first half of that century.

In the end, the historian of the movement, Lawrence Goodwyn, argues that the submersion of the Populist Party into the Democratic campaign of William Jennings Bryan in 1896 only confirmed the end of the Farmers' Alliance that had begun several years earlier with the crushing of local and state cooperatives by bankers, merchants, and buyers. That drained the democratic vitality out of the movement and, after Bryan's decisive defeat, left rural America convinced that the great cooperative campaign had failed because it "got into politics." In fact, politics was the only way to achieve the goals of the campaign, because the monetary system was stacked against farmers, and the farmers themselves were dependent for their existence upon commodity crops marketed nationally and internationally.[16] Nevertheless, as in the later Mexican case we will look at below, a core lesson is the danger of co-optation by a political party or government little attuned or sympathetic to the real issues facing farmers.

The power of the Alliance, like the power of the Grange before it, depended first of all on farmers democratically organized around practical goals and institutions, cooperatives above all. As

Goodwyn puts it, "In the world they created, they fulfilled the democratic promise in the only way it can be fulfilled—by people acting in democratic ways in their daily lives." That is what the cooperatives, and the Alliance halls that governed them, provided. As the cooperatives failed, Alliance organizers clung to the educational purposes of the order and its function as a community. But, notes Goodwyn, "a community cannot persist simply because some of its members have a strong conviction that it ought to persist. A community, even one seeing itself as a 'brotherhood' and 'sisterhood,' needs to have something fundamental to do, an organic purpose beyond 'fellowship' that reaffirms the community's need to continue its collective effort."[17] A community of farmers needs the likes of the cooperative to cement community, as well as provide a foundation for needed political action. These are important takeaways for the new farmers' movement, but ones that are scarcely audible as yet outside parts of the country where the heritage of the older movements can still be felt.

A similar movement in Mexico in the 1970s and '80s achieved considerable success for small growers before being overwhelmed by political upheaval. Starting in the mid-1970s peasant farmers raising coffee and small grains took control of their *ejidos* and of the Unions of Ejidos organized to enhance their economic power. The movement began to create sophisticated economic organizations, providing credit, crop insurance, collective buying power, and, above all, marketing leverage to peasant farmers. Aided by a new generation of leftist organizers with a dedication to grassroots control, these organizations created a national network with considerable lobbying power. Each organization charted its own course, building capacity to support peasant agriculture in the face of corrupt and inefficient government institutions. They established credit unions, assembled funds for crop insurance, and made their own deals with buyers, over the head of official agricultural purchasing arrangements. Many organizations hired a technical staff to provide the expertise necessary for the more sophisticated schemes they were developing. But power for the most part remained in the hands of the elected peasant leadership and the assemblies behind them, breaking the usual pattern of top-down decision making by NGOs, government advisers, and party functionaries typical of many developing countries.[18]

Mexico's accession to NAFTA in 1994, which eventually brought a flood of cheap maize into the country, and the subsequent Zapatista rebellion, divided the movement. Thanks to long-standing ties with government officials sympathetic to the movement, some organizations sided with the federal government, despite its program of neo-liberal economic reform and privatization of the ejidos. Others sided cautiously with the Zapatistas or stayed carefully on the sidelines. Some of the new institutions went under in the ensuing struggle, but others, particularly among coffee growers, have persisted, strengthened by the new Fair Trade movement. The divisions underlined the dangers of co-optation by government and the importance of maintaining genuinely independent, farmer-led organizations—ironically two of the central tenets of the original movement.

Most of the cooperative institutions created in the United States in the days of the Farmers' Alliance have atrophied or come to serve primarily the interests of the largest farmers or buyers. The Sunkist cooperative, for example, originally the Southern California Fruit Growers Exchange, came to dominate navel orange production in the nation, forming irrigation districts, raising biological controls for pests, and sponsoring a USDA research station to further their craft. But as land prices soared, farmer-owners became managers, then contracted out management to become absentee owners. And the co-op today is blamed for blatant disregard for the environment as its contractors spray and fumigate to preserve the picture-perfect image of its fruit.[19] In the West pork and beef producers' cooperatives sold out to the big packinghouses, only to find that they (and their children) had nowhere to sell their product but the oligopoly that has partitioned the country among a handful of regional monopolists.

Despite their failures and sometimes dubious heritage, these examples demonstrate the importance of organized efforts for both the economic and political power of farmers. Cooperative schemes put ownership squarely in the hands of producers, where it belongs. They also provide a forum where farmers can learn the character of the economic and political system they face and debate how to confront it. Cooperation can extend to most of the most important transactions in farming life—not just marketing but also equipment purchase and sharing, storage facilities, crop insurance, and credit. Cooperative enterprise remains the single most effective business form in the capitalist economy.[20] It can also facilitate access to health care and pension plans. And it can provide the economic power that translates into political influence, at least at the local level.

Where commodity production prevails, as in the grain and meat belt of the Midwest and Plains, the old fights are still relevant, though farmers' access to credit is no longer the central issue. Access to markets looms much larger, and upstart initiatives like the independent packinghouse and retail outlet spearheaded by Kansas rancher Mike Callicrate mirror the older cooperative movement. In Colorado and beyond, the Colorado Farmers Union carries on the cooperative heritage, today encouraging the development of cooperative food hubs to bring foods of all sorts to both local and more distant urban markets in the region. Callicrate has carried on his own policy campaigns, unsuccessfully suing the world's largest meatpacker, IPB, on anti-trust grounds and lobbying successive secretaries of agriculture to enforce the anti-trust laws against the meatpacking giants. Similarly, monopoly control of our seed stocks has become a hot-button issue for many food activists, prompting new legal vehicles for protecting varieties from exclusive copyright claims, and spawning new seed companies and new ventures in traditional plant breeding.

The political options pursued by farmers since the Populist days, on the other hand, have rarely been able to overcome the hegemony of corporate agribusiness and financial interests, not to mention the indifference of the general public. Nineteenth-century corporate control of the political system, solidified in the Republican campaign against the Populist insurgency, has only grown deeper. The country's willingness even to question the financial system was stifled until the collapse of 2008, and Congress today is busily undoing the meager reforms passed in its aftermath. As we saw in chapter 4, even the government's attempts to address the credit question systematically favored land concentration and the growth of agribusiness. And today prospects for mounting effective resistance to the agenda of the corporations seems small indeed.

## A Limited Politics for Today and Tomorrow

Our hope today, it appears, has to be local and focused on self-help, as Berry suggests. It will have to come from the sort of self-help organizations that the Grange and Farmers' Alliance tried unsuccessfully to create and that advocates like Callicrate and the Colorado Farmers Union are creating today. But such efforts are more locally focused than the nineteenth-century organizations, which were oriented always to a national market. Today many of us have the option of the local market, thanks in part to growing consumer sensibility about the foods we eat, and to the vacuum opened up by giant retail outlets sourcing internationally. That window, however, is a small one for the same reason: Those giants can offer food for all seasons virtually all the time, often at prices lower than we can meet.

The local food movement thus fights an uphill battle that will have to be bolstered with more and better farmer and consumer economic organization. Farmer-controlled food hubs promoting expanded market access is one hopeful sign. So is the growing urban farming movement. Finding ways to make farmers markets more inclusive of the whole population, from so-called market match programs for food stamp customers to concerted efforts to serve African American, Hispanic, Asian, and Native American farmers and buyers, should be on the agenda. So should local food-buying clubs and continued pressure on restaurants and grocers to buy local. We should also be supporting farmer education and internships, creating tool-sharing schemes, cooperative farm credit and crop insurance programs, cooperative storage facilities, and better ways to take care of one another and our families cooperatively.

Not that political advocacy is not still necessary. The regulatory state has grown increasingly demanding of farmers and food producers over the last decades. Though sometimes pursuing important aims, the new regulatory environment poses special challenges to small farmers, because legislators and bureaucrats alike prefer one-size-fits-all policies, however complex they might otherwise be. And lobbyists for agribusiness, corporate food handlers and processors, and the retail giants generally support that effort, while seeing to it that new regulations are ones they can afford. The result is that many of them are regulations small producers cannot afford. Older schemes for protecting farmers, like production quotas and price supports, have no hearing in Washington, thanks to the corporate lobbyists.[21] We need farmer-controlled advocacy organizations that are constantly alert to our particular needs as small farmers, engaged in creative adaptation to a changing agricultural reality.

We have few such organizations today in the United States. Most advocacy work on behalf of small farmers is in the hands of nonprofits, whose work may be well intentioned but is often ill informed and certainly not under the control of farmers themselves. And the same is true of many of the services that have been developed to provide for the economic interests of farmers, from farmers markets to lending agencies. As the federal government's largesse for funding such efforts fades, we will have to create our own institutions under our own control, as the Mexicans and our forebears in the Populist movement did.

And that is just as well, because the well-intentioned advocates, while doing the best they can for small farmers and local agriculture by their own lights, often end up as enablers for government programs that do not serve us well. Food safety is a case in point. Spurred on by neglect or malfeasance on the part of a few large packinghouses and processors, Congress passed sweeping new legislation in 2011 to regulate farm-level practices. Many states and localities also chimed in, as did corporate buyers. The result is a patchwork of regulations imposing often onerous burdens on farmers and small packing operations to regulate the tiny risk of contamination of food at the farm level. Two newcomers to the field, the National Sustainable Agriculture Coalition and the Farmers Market Coalition, contributed more to mobilizing farmers around the issue than any of the bigger, older farmers' groups. But some of the advocacy organizations, having won some and lost some in their battles for the best regulatory package available, have turned to educating farmers on food safety best practices. Sensible advice is mixed with the hysterical, and farmers are left to wonder when the first lawsuit or FDA SWAT team will attack their farm for some putative breach of protocol.

Genuine farmers' organizations, organized with grassroots participation at the forefront, might avoid some of the regulatory and educational excesses. They might provide the constant vigilance that we need to exercise in the current environment. Older farmers' organizations are rooted in a federated structure that enables superordinate organizations to meddle in local affairs, sometimes destroying the work of inspired local leaders. As we saw, even Oliver Kelley, founder of the Grange, ran afoul of the National Grange over cooperative buying and selling and political involvement. A recent ugly struggle between the California

State Grange and the national organization bitterly divided local organizations and discouraged many young farmers from deeper involvement. The Farm Bureaus are embedded in the American Farm Bureau Federation, created by bureaucrats at the USDA in 1919 to further the work of cooperative extension agents and promote "progressive" farming. Though local Farm Bureaus may mainly represent the interests of local farmers, state and national agendas are set at those levels and have routinely represented the interests of the largest farmers.

A few exemplary organizations exist today, like Oregon's Friends of Family Farmers, whose board and staff are dominated by working farmers. Through farmer-rancher listening sessions, they develop a legislative agenda and the support needed to back it. Their Oregon Pasture Network supports pasture-fed livestock production through networking and promotion. The sort of robust economic organization that underlies the Colorado Farmers Union is missing here. But the organization has an impact both on the legislative level and in promoting small farm products around the state. We will need to build on these and other models and experiences as we revitalize older organizations or create new ones crafted to our needs.

## Small Hopes and the Bigger Picture

Politics counts. Like the natural disasters that hit parts of the United States in the fall of 2017, it can sweep us away or alter drastically the terms on which we live. As we saw in chapter 5, governments and the social and economic forces they represent can make bad stewards of the land out of good farmers. They can drive us to marginal land or embroil us in debt, leaving us a stark choice between going under or abusing the land on which we rely. There aren't many things we can do about this, the perennial dream of a farmer-citizen republic notwithstanding. That's why the subsistence-first strategy advocated in chapter 3 is so important. To the extent we can provide for ourselves and our families, we become that much less vulnerable to the forces that would make us bad stewards of our land, or exiles from it. But for that we also need one another. We need community, and we need strong farmers' organizations.

The smaller, everyday political battles remain important. The growing demands of the bureaucratic state raise our expenses on a yearly basis. They make it more and more difficult just to farm as we struggle with compliance over bogus food safety measures or certifications or water management requirements. The propensity of regulators to create one-size-fits-all rules, plus the design of regulations to manage (and appease) the biggest operators, make regulations all the more questionable for small farmers and food producers. The tensions the regulatory regimes create are all the more difficult to manage because some of these requirements are reasonable, while others are clearly over the top or a waste of time. How do we oppose regulations that answer some real social need, even if badly, when many of our customers and neighbors assume that the state is just trying to help? How do we mount a real opposition when most producers have already given up hope of influencing these decisions, which they resignedly treat as just another cost of doing business?

In the long term we may not have to face these questions. The next financial collapse, or the next, may end what little we have of the nanny state, leaving only the police state. And that is already notably weak when it comes to the reach of regulatory enforcement. Last I heard, California's Department of Food and Agriculture (CDFA) dairy enforcement division for Northern California, located in Oakland, had just four inspectors and a secretary. After a brief flurry of raids on illegal backyard dairies throughout the state a few years ago, provoking widespread public outrage, and efforts at reform blocked by big dairy, CDFA backed off in an ignominious retreat. No dairy laws were changed, but a measure of tolerance was achieved. Let us hope that such ephemeral struggles are the worst we can expect in coming years.

In the long haul, though, history offers greater comforts. Sociologist Barrington Moore ends his study of "lord and peasant in the making of the modern world" with the observation that "the process of modernization begins with peasant revolutions that fail. It culminates during the twentieth century with peasant revolutions that succeed. No longer is it possible to take seriously the view that the peasant is an 'object of history,' a form of social life over which historical changes pass but which contributes nothing to the impetus of these changes. For those who savor historical irony it is indeed curious that the peasant in the modern era has been as much an agent of revolution as the machine, that he has come into his own as an effective historical actor along with the conquests of the machine." Though Moore shares the pessimistic appraisal of the outcome of these revolutions that we read earlier in the words of James Scott, he goes on to credit "the village" with "those half-conscious standards by which men have judged and condemned modern industrial civilization, the background from which they have formed their conceptions of justice and injustice." And he argues that "the wellsprings of human freedom lie not only where Marx saw them, in the aspirations of classes about to take power, but perhaps even more in the dying wail of a class over whom the wave of progress is about to roll"—that is, the peasantry.[22] As it turns out, however, the peasantry, and the small farmer generally, may have been rolled over, but they have scarcely died out.

The depressing portrait of defeat even in victory that students of peasant rebellion have painted, from Moore to Scott, comes up short of the truth. In the former Soviet Union, Stalin replaced a mainly peasant agriculture with massive state and collective farms built on "modern" lines in a campaign that cost as many as six million peasant lives. Yet by the 1980s, between a fourth and half of all food on Soviet tables was being produced on small plots averaging half an acre, and these household plots and small farms persist as major producers today.[23]

Similarly, in China the peasant household reemerged as the dominant unit of production in the wake of economic reforms under Mao's successor, Deng Xiaoping. Writing in 1987, Elisabeth Croll notes that "ten years ago any analysis of the rural production system would have focused on the commune, the production brigade and production teams and would only have touched on the peasant household. Now . . . an analysis of the new rural production system is centered on an understanding of the peasant household economy and its significance as an important sociopolitical unit." From a system in which "production was planned and managed by the production team which decided the crops, timetabled the production process, allocated production tasks to its members and disposed of the product," China has returned to the ancient system in which the household handles all these tasks. More impressive still, the countryside led the wave of prosperity that followed Deng's reforms. Fifteen years later another analyst notes the concern among Chinese officials about how to integrate peasants into a thriving internal economy, commenting that, "The peasant question continually returns no matter how many times state authorities and intellectuals declare that it has been definitively resolved."[24]

The "end of the peasantry" has been announced repeatedly and was and is, indeed, an article of faith for liberal scholars, American policy makers, Marxists, conservative political scientists, and radical sociologists alike. "Modernization," it was thought, would put an end to traditional agriculture with its basis in strategies of subsistence first, diversified farming, and rural community. That way of life was, at any rate, nasty, brutish, and short, advocates of modernity have insisted, and best exchanged for a good job in the city, or, better yet, in the agriculture department of a major university. Alas, neither the expectation nor its portrait of rural life have proven sustainable.

Traditional agriculture—inventive, artful, endlessly experimental, if also doggedly persistent, hardscrabble, family- and community-bound—is here to stay. It will outlast the industrial alternative, as it has outperformed it in both Russia and China. And it will sustain humanity, provided we take its many lessons to heart, through whatever trials our civilization faces in the years to come.

# ACKNOWLEDGMENTS

This book is the product of fifty years of reading and thinking about the ways in which peoples around the world have subsisted on the land, twelve years of intermittent research in Mexico and El Salvador, and over a decade of farming—none of which gives its author any special authority, but all of which, put together, entails a lot of debts, both personal and intellectual. I will start with the personal, as these are always the most important.

I owe special thanks to my two principal readers. My daughter Julian Foley discovered while in journalism school a decided talent for editing, and she has saved me from more than a few confusions of organization and argument and many a stylistic peccadillo. My wife, Sara Grusky, besides detailed editorial advice, has forced consideration and reconsideration of important parts of the work in progress, making it decidedly the conversation that it pretends to be. Without her engagement with the substance of the work, it would not be what it is—which is not to deny that is it "my" book, as she insists, for which I must take full responsibility.

I also want to thank my editors at Chelsea Green, first Michael Metivier, then Ben Watson, for their enthusiasm for this book and their continuing willingness to help, and to the staff who helped put this all together. Chelsea Green was my first choice of a publisher because the press, far more than any other today, has produced the kinds of books we rely upon, not only as we build a renewed farming practice and culture, but also as we think through our place in these troubled times.

To my shame and loss, I am not the sort to seek out mentorship or even advice, either as a scholar or a farmer, but there are nevertheless personal debts to people with whom I have worked that must be acknowledged. One is to Luis Hernández Navarro, organizer and journalist, who tolerated my only slowly improving Spanish while giving me access to his many contacts in the Mexican peasant movement of the 1980s and '90s. Luis insisted that this gringo researcher give back at least some of what he took, and though I have never felt I honored that obligation, I hope this book returns some of the insights I gained through his help.

Another more contemporary debt is to Ruthie King, the dynamic young director of the School of Adaptive Agriculture, the farmer training project that I helped found a few years ago. Without Ruthie's curiosity and knowledge about all things livestock, and her ability to bring speakers and workshops to the school around her passions, I would know far less than I do about holistic management, pasture-cropping, animal welfare, and carbon farming. And thanks, too, to my daughter Allegra, whose decision to pursue first an internship on a farm, then the apprenticeship program at the Center for Agroecology and Sustainable Food Systems at Santa Cruz, led directly to my decision to take up farming.

My intellectual debts are too many to enumerate, so I'll just list some of the biggest. To Karl Polanyi's *The Great Transformation*, barely referenced here, which I first read in the 1970s, I owe my appreciation for the contrast between the sorts of societies that went before the triumph of capitalism and our current economic arrangements. I also owe Polanyi's work an enormous amount for his critique of the neoliberal nostrums that took hold in the 1980s, wreaking havoc on the world of so-called developing countries that I was beginning to study about that time, not to mention our own. Equally important, and unrecognized here, has been the early work of Ivan Illich, which exposed the flaws in the modernist utopia that technological progress was supposedly bringing us.

# ACKNOWLEDGMENTS

I read James Scott's *The Moral Economy of the Peasant* about the time I was thinking of returning to graduate school. There I found a political scientist working on themes that I had been fascinated by for a decade or more, and I was convinced that I could do similar work as a political scientist. Alas, Scott has always owed far more to the anthropologists and historians than to the narrow methods and preoccupations of political science, which I never managed to surmount with his finesse during my twenty-year career in the field. But his subsequent work has kept me inspired and, as those who attend to the endnotes will have seen, contributed enormously to this book.

Wendell Berry's long-running critique of industrial agriculture and its foundation in our culture of neglect (for the land, for its people, for consequences in general) has inspired all of us who espouse a genuinely sustainable agriculture. But his persistent effort to ground farming in a culture of respect for the land and in a perduring community based on the "land economies" is the deepest contribution we could ever want for thinking through how we are to live in the coming years. This book is deeply indebted to his thought. I only hope that I have done him justice.

This book would not have been possible without the decade-long engagement with this place that Sara and I and our family have undertaken. The experience of rehabilitating an old farm on beautiful, if marginal, land and creating viable market garden and small goat dairy enterprises has reshaped our lives and thinking. We came to settle and farm in Willits, California, in part to participate in a movement that anticipated many of the features of what became the Transition Towns initiative. Despite the spotty fulfillment of the dream of economic localization that helped draw us, a vibrant local food movement has grown up here, providing both fellowship and hope. And the relationships we have built, particularly with younger farmers and aspiring farmers, have sustained us and inspired much that I have written here.

Finally, I wish to thank my wife, Sara, once again. Her encouragement, engagement, and understanding for what it takes to launch and finish a project like this made this book. Our intellectual journey together has deepened with this project, but she will probably never know how indebted to her I am for how I approach and understand the world thanks to our years as partners.

# NOTES

**Chapter 1: Farming in the Ruins of the Twentieth Century**

1. John Michael Greer is the latest to explore the decline of our own civilization in the light of past experiences. See his *The Long Descent: A User's Guide to the End of the Industrial Age* (Gabriola Island, BC: New Society, 2008).
2. See Andrew Kimbrell, editor, *Fatal Harvest: The Tragedy of Industrial Agriculture* (Washington, DC, Covelo, CA, and London: Island Press, 2002); and Wendell Berry's indispensable *The Unsettling of America: Culture and Agriculture* (San Francisco: Sierra Club Books, 1977).
3. Satoshi Katazawa, "Common Misconceptions About Science II: Life Expectancy," *Psychology Today*, November 20, 2008, https://www.psychologytoday.com/us/blog/the-scientific-fundamentalist/200811/common-misconceptions-about-science-ii-life-expectancy. The research on historical and contemporary hunter-gatherer cultures is cited in Anthony A. Volk and Jeremy Atkinson, "Is Child Death the Crucible of Human Evolution?," *Journal of Cultural and Evolutionary Psychology* (December 2008), Proceedings of the 2nd Annual Meeting of the Northeastern Evolutionary Psychology Society.
4. The classic account is Wendell Berry's *The Unsettling of America*. Few people seem to be aware of the efforts of the largest corporations to destroy local autonomy. Big meat-packers, for example, bought up small, sometimes cooperatively owned packing houses, dismantled them, then portioned out the country among themselves to make each of the giant packers monopolists in their own region. In tandem with that process, supermarket chains eliminated skilled butchers from their workforces, relying on packaged meat from the big packing houses.
5. That is British Petroleum's figure, derived from a number of governmental and nongovernmental sources and presented on one of the more accessible sites (http://www.bp.com/en/global/corporate/energy-economics/statistical-review-of-world-energy/oil/oil-reserves.html). BP notes that

estimates of "proved reserves" have grown with new discoveries and may be expected to do so in the future. And their figures are based on current use, which has also been growing. Both the rate of new discovery and that of demand have grown more slowly than in the past. In any case the end is the end, and prices will certainly rise well before the actual exhaustion of oil production, as Richard Heinberg argued in what is still one of the most relevant books on the end of the Age of Oil, *The Party's Over: Oil, War and the Fate of Industrial Societies*, revised and updated edition (Gabriola Island, BC: New Society, 2005). His ongoing blog at the Post Carbon Institute tracks the ups and downs of the oil economy: http://www.postcarbon.org.

6. George Monbiot, "We're Treating the Soil Like Dirt," *The Guardian*, March 25, 2015, https://www.theguardian.com/commentisfree/2015/mar/25/treating-soil-like-dirt-fatal-mistake-human-life.
7. As early as 1978, Warren Johnson had some perspicacious thoughts about the effect of energy decline on American agriculture. We may quibble with the particulars, but the picture he paints remains relevant today. See Johnson, *Muddling Toward Frugality* (San Francisco: Sierra Club Books, 1978), 74–78.
8. A call for submissions to the *New Famer's Almanac*, vol. 3: https://thegreenhorns.wordpress.com/2016/01/20/reminder-call-for-submissions-to-the-almanac.
9. This is the message of Ben Falk's *The Resilient Farm and Homestead* (White River Junction, VT: Chelsea Green, 2013).
10. Wendell Berry, "The Making of a Marginal Farm," reprinted in *The World-Ending Fire: The Essential Wendell Berry* (Allen Lane, 2017), 44.
11. "What Is Resilience?" *Resilience*, http://www.resilience.org/about-resilience/#resilience.
12. Berry, "The Making of a Marginal Farm," 45.

**Chapter Two: A Short, Unhappy History of Business Advice for Farmers**

1. Deborah Fitzgerald, *Every Farm a Factory: The Industrial Ideal in American Agriculture* (New Haven and London: Yale University Press, 2003).
2. Wendell Berry, *The Unsettling of America: Culture and Agriculture* (San Francisco: Sierra Club Books, 1977), vii–viii.
3. Wendell Berry, "Money Versus Goods," in *What Matters? Economics for a Renewed Commonwealth* (Berkeley: Counterpoint, 2010), 24.
4. CRAFT is a collection of farmer-organized local or regional networks for training new farmers. With examples around the country, the movement has posted various resources online at: http://www.craftfarmer.org.
5. One of the most popular recent books on the subject, Richard Wiswall's *The Organic Farmer's Business Handbook* (White River Junction, VT: Chelsea

Green, 2009), comes from a market gardener—a farmer, that is, who appreciates a diverse cropping and marketing strategy. But Wiswall echoes the cost-accounting advice of the most simple-minded ag economists of the 1920s. Let's do a "crop enterprise budget," crop by crop, Wiswall advises, so we know which crops are worth growing, which not. Presumably, Wiswall is aware that a diversified strategy can protect the farmer against the failure of a given crop. He is not headed down the path of monoculture. But his economic logic is the same that brought us to the current state of American agriculture. And there is no mention in Wiswall of the subsistence strategies—the woodlot, flock of chickens, couple of pigs, dairy cow or goats, and cooperative labor arrangements—which many of the new farmers have adopted for reasons that their would-be advisers inevitably dismiss as "romantic." If it can't be monetized, it can't be counted and doesn't then count.

**Chapter Three: Subsistence First!**

1. John Michael Greer, *Dark Age America: Climate Change, Cultural Collapse, and the Hard Future Ahead* (Gabriola Island, BC: New Society, 2016), 106.
2. Kevin Healy, *Llamas, Weavings, and Organic Chocolate: Multicultural Grassroots Development in the Andes and Amazon of Bolivia* (South Bend, IN: University of Notre Dame Press, 2001).
3. Edgar Anderson, *Plants, Man and Life* (Berkeley, Los Angeles, London: University of California Press, 1971), 136–42.
4. James C. Scott, *The Art of Not Being Governed: An Anarchist History of Upland Southeast Asia* (New Haven and London: Yale University Press, 2010), 195.
5. Paul Richards, *Indigenous Agricultural Revolution* (Boulder, CO: Westview Press, 1985), 66–70.
6. Richards, *Indigenous Agricultural Revolution*, 7.
7. Gary Paul Nabhan, *Enduring Seeds: Native American Agriculture and Wild Plant Conservation* (Tucson: University of Arizona Press, 1989), 31–37.
8. Nabhan, *Enduring Seeds,* 38, 39.
9. Richards, *Indigenous Agricultural Revolution*, 98–101, 77.
10. James C. Scott, *Seeing Like a State: How Certain Schemes to Improve the Human Condition Have Failed* (New Haven and London: Yale University Press, 1998), 304.
11. Gene C. Wilken, *Good Farmers: Traditional Agricultural Resource Management in Mexico and Central America* (Berkeley, Los Angeles, London: University of California Press, 1987), 225.
12. Chris Smaje, "Three Stories, Many Variables: The Case for Small Farm Productivity," *Statistics Views*, March 5, 2015, http://www.statisticsviews

.com/details/feature/7553541/Three-stories-many-variables-the-case-of-small-farm-productivity.html.

13. Jean-Martin Fortier, *The Market Gardener: A Successful Grower's Handbook for Small-Scale Organic Farming* (Gabriola Island, BC: New Society, 2014).
14. Eliot Coleman, *The New Organic Grower: A Master's Manual of Tools and Techniques for the Home and Market Gardener*, revised and expanded edition (White River Junction, VT: Chelsea Green, 1995).
15. Ben Hartman, *The Lean Farm* (White River Junction, VT: Chelsea Green, 2015).
16. For a thorough, if a bit dated, article on the Kaisers' Singing Frogs Farm, see Todd Oppenheimer, "The Drought Fighter," *Craftsmanship Quarterly*, January 2015, https://craftsmanship.net/drought-fighters.
17. Perrine Hervé-Gruyer and Charles Hervé-Gruyer, *Miraculous Abundance: One Quarter Acre, Two French Farmers, and Enough Food to Feed the World* (White River Junction, VT: Chelsea Green, 2016).
18. See Laura Lengnick, *Resilient Agriculture: Cultivating Food Systems for a Changing Climate* (Gabriola Island, BC: New Society, 2015), 207–21.
19. Courtney White, *Grass, Soil, Hope: A Journey Through Carbon Country* (White River Junction, VT: Chelsea Green, 2013).
20. Mark Shepard, *Restoration Agriculture: Real-World Permaculture for Farmers* (Austin, TX: Acres USA, 2013).
21. Carol Deppe, *The Resilient Gardener: Food Production and Self-Reliance in Uncertain Times* (White River Junction, VT: Chelsea Green, 2010); *The Tao of Vegetable Gardening: Cultivating Tomatoes, Greens, Peas, Beans, Squash, Joy, and Serenity* (White River Junction, VT: Chelsea Green, 2015); *Breed Your Own Vegetable Varieties: The Gardener's and Farmer's Guide to Plant Breeding and Seed Saving* (White River Junction, VT: Chelsea Green, 2000).
22. Will Bonsall, *Will Bonsall's Essential Guide to Radical, Self-Reliant Gardening* (White River Junction, VT: Chelsea Green, 2015), 99.
23. Nabhan, *Enduring Seeds*, 96.
24. Michael Phillips, *The Holistic Orchard: Tree Fruits and Berries the Biological Way* (White River Junction, VT: Chelsea Green, 2012).
25. Richard Heinberg and David Findley, *Our Renewable Future: Laying the Path for One Hundred Percent Clean Energy* (Washington, DC: Island Press, 2016). But they also show that we can expect to have much less energy to use than the Age of Oil has provided.
26. John Michael Greer, *The EcoTechnic Future: Envisioning a Post-Peak World* (Gabriola Island, BC: New Society, 2009), 154–58.

27. John Michael Greer's discussion in *Dark Age America*, 160–68, is the only considered analysis of the future of the internet in an age of energy contraction I've found.

**Chapter Four: Land for the Tiller**

1. Simon Fairlie, *Low Impact Development: Planning and People in a Sustainable Countryside* (Charlbury, Oxfordshire, U.K.: Jon Carpenter, 1996).
2. Lawrence Goodwyn, *The Populist Moment: A Short History of the Agrarian Revolt in America* (Oxford, London, New York: Oxford University Press, 1978), 269, 297–98.
3. Julie Guthman, *Agrarian Dreams: The Paradox of Organic Farming in California* (Berkeley, Los Angeles, London: University of California Press, 2004).
4. Joel K. Bourne Jr., *The End of Plenty: The Race to Feed a Crowded World* (New York: W. W. Norton, 2015).
5. Simon Fairlie, "The Tragedy of the Tragedy of the Commons," *Dark Mountain* 1 (Summer 2010), 178–200.
6. Elinor Ostrom, *Governing the Commons: The Evolution of Institutions for Collective Action* (New York and Cambridge, U.K.: Cambridge University Press, 1990).
7. Fairlie, "The Tragedy of the Tragedy of the Commons."
8. Andro Linklater, *Owning the Earth: The Transforming History of Land Ownership* (New York, London, New Delhi, Sydney: Bloomsbury, 2013), 5.
9. Linklater, *Owning the Earth*, 31.
10. Linklater, *Owning the Earth*, 35.
11. Linklater, *Owning the Earth*, 209.
12. Paul Wallace Gates, "The Homestead Law in an Incongruous Land System," in *The People's Land: A Reader on Land Reform in the United States*, edited by Peter Barnes (Emmaus, PA: Rodale Press, 1975), 7–11.
13. David Graeber, *Debt: The First 5000 Years* (Brooklyn, NY: Melville House, 2011), 64–65, 81–82.
14. Fairlie, "The Tragedy of the Tragedy of the Commons," 191.
15. James Mahoney, *The Legacies of Liberalism: Path Dependence and Political Regimes in Central America* (Baltimore: Johns Hopkins University Press, 2001).
16. Angus MacDonald, "The Family Farm Is the Most Efficient Unit of Production," in *The People's Land*, 86.
17. George Monbiot, "Peasant Farmers Offer the Best Chance of Feeding the World. So Why Do We Treat Them with Contempt?" *The Guardian*, June 10, 2008.

18. Chris Newman, "Yes, Organic Farming Will Kill Us All," *NewCo Shift*, January 12, 2017, https://shift.newco.co/yes-organic-farming-will-kill-us-all-12d900979cf2.
19. See his blog, *Small Farm Future*, http://smallfarmfuture.org.uk, especially his posts on a "neo-peasant" Wessex, at https://smallfarmfuture.org.uk/category/neo-peasant-wessex.
20. As Mexico moved to "modernize" its economy on neo-liberal lines under the administration of Carlos Salinas de Gortari in the early 1990s, Mexico's secretary of state declared that "Mexico has too many peasants." He was unconsciously echoing American planners in the 1960s who maintained that the United States suffered from a surplus of farmers.
21. Fairlie, "The Tragedy of the Tragedy of the Commons," 184; Marc Bloch, *French Rural History: An Essay on Its Basic Characteristics* (Berkeley and Los Angeles: University of California Press, 1966), 44.
22. Alan Mayhew, *Rural Settlement and Farming in Germany* (London: B. T. Batsford, 1973).
23. Bloch, *French Rural History*, 60–62.
24. James C. Scott, *The Moral Economy of the Peasant: Rebellion and Subsistence in Southeast Asia* (New Haven and London: Yale University Press, 1976).
25. M. Kat Anderson, *Tending the Wild: Native American Knowledge and the Management of California's Natural Resources* (Berkeley, Los Angeles, London: University of California Press, 2005).
26. Brian Donahue, *Reclaiming the Commons: Community Farms and Forests in a New England Town* (New Haven and London: Yale University Press, 1999).
27. Mike Madison, *Fruitful Labor: The Ecology, Economy, and Practice of a Family Farm* (White River Junction, VT: Chelsea Green, 2018).
28. "What We Do," Slow Money, https://slowmoney.org/about/our-work/what-we-do.
29. The details can be found by accessing the articles of incorporation and bylaws on the member application page, https://www.ourtable.us/application-form.html.
30. National Incubator Farm Training Initiative, https://nesfp.org/food-systems/national-incubator-farm-training-initiative.
31. Donahue, *Reclaiming the Commons*, 286.
32. For a good survey of such methods, see Tom Atlee, *The Tao of Democracy: Using Co-Intelligence to Create a World That Works for All* (Cranston, RI: Writers Collective, 2003).
33. Cooperatives offer a contemporary framework for such organization. But the one person/one vote and majority rule features of the cooperative

model mean that a simple majority can simply sell off assets if that prospect proves attractive. Many agricultural cooperatives of the past have met this fate, as older owners cashed out on lucrative opportunities offered by the growing monopolies in agricultural processing and marketing. The hybrid model pioneered by Poudre Valley Community Farms may provide extra assurance that hard-won land tenure arrangements will not evaporate with a change of board or member sentiment, but there are no absolute guarantees under modern Western property law.

**Chapter Five: Soil, Civilization, and Resilient Farmers Through the Centuries**

1. Modern Western attention to the issue of soil degradation began with George Perkins Marsh's 1864 *Man and Nature; or, Physical Geography as Modified by Human Action* (New York: Charles Scribner, 1864). In 1953 W. C. Lowdermilk published the results of his field investigations funded by the US Department of Agriculture in *Conquest of the Land Through 7,000 Years*, US Department of Agriculture, Soil Conservation Service, Agriculture Information Bulletin 99 (Washington, DC: Government Printing Office, 1953). In 1956 the Wenner-Gren Foundation held an international anthropological symposium on the topic chaired by Carl Sauer, Marston Bates, and Lewis Mumford. The results were published as *Man's Role in Shaping the Face of the Earth*, edited by William L. Thomas Jr. (Chicago: University of Chicago Press, 1956). Since the 1952 publication of E. Hyams, *Soil and Civilization* (London and New York: Thames and Hudson), a number of writers have argued that soil degradation was the direct cause of the collapse of many ancient civilizations. David R. Montgomery, in *Dirt: The Erosion of Civilizations* (Berkeley, Los Angeles, London: University of California Press, 2007), lays out the most recent historical evidence and modern patterns of soil degradation.
2. Montgomery, *Dirt*, 74–80, 84.
3. Ester Boserup, *The Conditions of Agricultural Growth: The Economics of Agrarian Change Under Population Pressure* (Chicago: Aldine, 1965), 28–31, 44–48; Marshal Sahlins, *Stone Age Economics* (Chicago: Aldine, 1972), chapter 1.
4. David Christian, *Maps of Time: An Introduction to Big History* (Berkeley, Los Angeles, London: University of California Press, 2004), 232.
5. Roy A. Rappaport, *Pigs for the Ancestors: Ritual in the Ecology of a New Guinea People*, revised edition (New Haven: Yale University Press, 1984).
6. Boserup, *Conditions of Agricultural Growth*, 73.
7. James C. Scott, *Against the Grain: A Deep History of the Earliest States* (New Haven and London: Yale University Press, 2017), chapter 5.

8. David Graeber, *Debt: The First 5,000 Years* (Brooklyn, NY: Melville House, 2011).
9. Montgomery, *Dirt*, 166–67.
10. Charles C. Mann, *1493: Uncovering the New World Columbus Created* (New York: Knopf, 2011), 167–77.
11. Pierre Clastres, *Society Against the State: Essays in Political Anthropology* (New York: Zone Books, 1987), 188.
12. James C. Scott, *The Art of Not Being Governed: An Anarchist History of Upland Southeast Asia* (New Haven and London: Yale University Press, 2010).
13. George M. Foster, "Peasant Society and the Image of the Limited Good," *American Anthropologist New Series* 67, no. 2 (April 1965), 293–315.
14. Chris Arsenault, "Only 60 Years of Farming Left if Soil Degradation Continues," *Scientific American*, 2014, https://www.scientificamerican.com/article/only-60-years-of-farming-left-if-soil-degradation-continues.
15. Gary Paul Nabhan, *Growing Food in a Hotter, Drier Land: Lessons from Desert Farmers on Adapting to Climate Uncertainty* (White River Junction, VT: Chelsea Green, 2013), 158–60.
16. Most of this discussion relies on Gene C. Wilken, *Good Farmers: Traditional Agricultural Resource Management in Mexico and Central America* (Berkeley, Los Angeles, London: University of California Press, 1987), 104–26.
17. Paul Richards, *Indigenous Agricultural Revolution* (Boulder, CO: Westview Press, 1985), 55–62.
18. Charles C. Mann, *1491: New Revelations of the Americas Before Columbus* (New York: Knopf, 2005), 302–6.
19. Mann, *1491*, 306–11.
20. Nabhan, *Growing Food*, 156.
21. Montgomery, *Dirt*, 121–29.
22. Wilken, *Good Farmers*, 48–49.
23. Malcolm Thick, *The Neat House Gardens: Early Market Gardening Around London* (Totnes, Devon, U.K.: Prospect Books, 1998), 101–2.
24. Joseph Courtois-Gérard, *Practical Handbook of Market Gardening*, 6th edition, translated by Carol Cox (Willits, CA: self-published by translator, November 2017).
25. F. H. King, *Farmers of Forty Centuries: Organic Farming in China, Korea, and Japan* (Mineola, NY: Dover Publications, 2004 [1911]), 189–99.
26. Montgomery, *Dirt*, 40–43.
27. Wilken, *Good Farmers*, 70–82. Wilken notes that some of these diversions rely on long-standing erosion in neighboring uplands.

28. Nabhan, *Growing Food,* 119–20.
29. King, *Farmers of Forty Centuries*, 109–17, 180–89.
30. K. L. Sahrawat, "Fertility and Organic Matter in Submerged Rice Soils," *Current Science* 88, no. 5 (March 10, 2005), 735–39.
31. Nabhan, *Growing Food*, 160–65.
32. Cynthia Graber, "Farming Like the Incas," Smithsonian.com, September 6, 2011, https://www.smithsonianmag.com/history/farming-like-the-incas-70263217.
33. Paul and Elizabeth Kaiser, "Our Farming Model," Singing Frogs Farm, http://singingfrogsfarm.com/our-farming-model.html.
34. Bob Canard, Green String Farm, http://greenstringfarm.com/about.
35. Courtney White, *Grass, Soil, and Hope: A Journey Through Carbon Country* (White River Junction, VT: Chelsea Green, 2013); David R. Montgomery, *Growing a Revolution: Bringing Our Soil Back to Life* (New York and London: W. W. Norton, 2017), chapter 11; and David R. Montgomery and Anne Biklé, *The Hidden Half of Nature: The Microbial Roots of Life and Health* (New York and London: W. W. Norton, 2016).
36. Montgomery, *Growing a Revolution*, 68, 94.
37. Montgomery, *Growing A Revolution*, 20.
38. See The Land Institute, https://landinstitute.org/our-work/perennial-crops.
39. J. Russell Smith, *Tree Crops: A Permanent Agriculture* (Washington, DC, and Covelo, CA: Island Press, 1950 [1929]).
40. Eliot Coleman, *The New Organic Grower: A Master's Manual of Tools and Techniques for the Home and Market Gardener*, revised and expanded edition (White River Junction, VT: Chelsea Green, 1995), 94–118.
41. Will Bonsall, *Will Bonsall's Essential Guide to Radical, Self-Reliant Gardening* (White River Junction, VT: Chelsea Green, 2015), 59–62, 48–54. Michael Phillips also recommends spreading woody mulch around the drip line of fruit trees, *The Apple Grower: A Guide for the Organic Orchardist*, 2nd edition (White River Junction, VT: Chelsea Green, 2005).
42. For other examples see Laura Lengnick, *Resilient Agriculture: Cultivating Food Systems for a Changing Climate* (Gabriola Island, BC: New Society, 2015).
43. Coleman, *New Organic Grower*, 71–81.
44. See, for instance, Montgomery and Biklé, *Hidden Half of Nature*; Didi Pershouse's eclectic *The Ecology of Care: Medicine, Agriculture, Money, and the Quiet Power of Human and Microbial Communities* (Thetford Center, VT: Mycelium Books, 2016); and Stephen Harrod Buhner, *The Lost Language of Plants: The Ecological Importance of Plant Medicines to Life on Earth* (White River Junction, VT: Chelsea Green, 2002).

**Chapter Six: Resourceful Farmers**

1. It may be worth keeping up with the best scientific predictions of the regional impacts of climate change. For the United States, the periodic National Climate Assessment is a cautious look at the future. Provided the federal government continues to fund these efforts, they are a resource for farmers looking to understand their future. The latest NCA utilizes data and scientific findings through 2016. The report is available at: https://science2017.globalchange.gov.
2. Gary Paul Nabhan, *Enduring Seeds: Native American Agriculture and Wild Plant Conservation* (Tucson: University of Arizona Press,. 1989), 72, 75–77.
3. Nabhan, *Enduring Seeds*, 96.
4. Carol Deppe, *Breed Your Own Vegetable Varieties: The Gardener's and Farmer's Guide to Plant Breeding and Seed Saving* (White River Junction, VT: Chelsea Green, 1993, 2000); Will Bonsall, *Will Bonsall's Essential Guide to Radical, Self-Reliant Gardening* (White River Junction, VT: Chelsea Green, 2015), 73–100.
5. Gary Paul Nabhan has a useful introduction to the problems of "milder winters, earlier springs, hotter summers, and fruitless falls" in chapter 7 of his *Growing Food in a Hotter, Drier Land: Lessons from Desert Farmers on Adapting to Climate Uncertainty* (White River Junction, VT: Chelsea Green, 2013). The chapter includes a table of common fruit and nut species and varieties, and the chill hours each requires.
6. Joseph Courtois-Gérard, *Practical Handbook of Market Gardening*, 6th edition, translated by Carol Cox (Willits, CA: self-published by translator, November 2017). Chadwick was the founder of the horticulture training program on the University of California–Santa Cruz campus that evolved into the Center for Agroecology and Sustainable Food Systems' apprenticeship program. He went on to inspire a generation or two of market gardeners in Northern California and beyond. Chadwick, who had worked with Rudolf Steiner, also brought biodynamic thinking to many of his students. The best introduction to Chadwick's work and influence can be found at http://alan-chadwick.org/index.html. Eliot Coleman is also beholden to the tradition of the French market gardeners, and John Jeavons's biointensive system is a reflection of his own apprenticeship with Chadwick.
7. Kris De Decker, "Fruit Walls: Urban Farming in the 1600s," *Low-Tech Magazine*, December 24, 2015, http://www.lowtechmagazine.com/2015/12/fruit-walls-urban-farming.html.
8. One comment to the article referenced above wrote, "Plant a peach tree on the southeast side of a pond (about 1⁄16 acre in size = 50 × 50 feet

square and at least 3 feet deep) then mulch the tree with stones 8 inches thick. (Apply rock mulch from the trunk to the 'drip line' = the end of the farthest branch). The combination of rocks and water can raise canopy temperatures by 5°F, just enough to prevent frost damage to blossoms. Planting the peach tree near a large boulder, natural cliff, or fruit wall 8 to 10 feet high provides additional frost protection. A south-facing cliff, small pond, and rock mulch can raise canopy temperatures about 18°F. This is how we grow tender fruits in the Austrian Alps. This horticultural technology dates back to the Renaissance. Greek and Roman agronomists also wrote about the warming power of rocks. I have visited commercial orchards in Northern India where apple trees were planted among boulders as big as cars and mobile homes. The trees thrived in the protected micro-climate provided by the huge rocks which not only provide heat but also keep the soil constantly moist." Eric Koperek blogs on alternative agriculture at https://worldagriculturesolutions.com.

9. Carole Crews, *Clay Culture: Plasters, Paints and Preservation* (Ranchos de Taos, NM: Gourmet Adobe Press, 2010), 76–77.
10. Gene C. Wilken, *Good Farmers: Traditional Agricultural Resource Management in Mexico and Central America* (Berkeley, Los Angeles, London: University of California Press, 1987), 142–44.
11. David R. Montgomery, *Growing a Revolution: Bringing Our Soil Back to Life* (New York and London: W. W. Norton, 2017), 134–35.
12. Paul Richards, *Indigenous Agricultural Revolution* (Boulder, CO: Westview Press, 1985), 64–71.
13. Eliot Coleman explains the process in *Four-Season Harvest: Organic Vegetables from Your Home Garden All Year Long* (White River Junction, VT: Chelsea Green, 1999), 139–41.
14. Mike and Nancy Bubel, *Root Cellaring: Natural Cold Storage of Fruits and Vegetables*, 2nd edition (North Adams, MA: Storey Publishing, 1991).
15. Sandor Ellix Katz, *The Art of Fermentation* (White River Junction, VT: Chelsea Green, 2012).
16. John Michael Greer, *The EcoTechnic Future: Envisioning a Post-Peak World* (Gabriola Island, BC: New Society, 2009), 154–58, 70–74.
17. *Global Water Security: Intelligence Community Assessment*, 2012, available at https://www.dni.gov/files/documents/Special%20Report_ICA%20Global%20Water%20Security.pdf.
18. Wilken, *Good Farmers*, 214–15.
19. Wilken, *Good Farmers*, 219–22.
20. Richards, *Indigenous Agricultural Revolution*, 74–83.

21. Karl A. Wittfogel, *Oriental Despotism: A Comparative Study of Total Power* (New Haven: Yale University Press, 1957).
22. WaterHistory.org, "Water History—Aguadas, Cenotes, and Chultuns," http://waterhistory.org/histories/aguadas.
23. C. Kenneth Pearse, "Qanats in the Old World: Horizontal Wells in the New," *Journal of Range Management* 26, no. 5 (September 1973), 320–21; and Dale R. Lightfoot, "Syrian Qanat Romani" (http://waterhistory.org/histories/syria).
24. Wilkens, *Good Farmers*, 198–211; Pearse, "Qanats in the Old World."
25. "Plumbing," Wikipedia, https://en.wikipedia.org/wiki/Plumbing#Water_pipes.
26. Ravindra Krishnamurthy, "Bamboo Drip Irrigation," *Permaculture News*, February 28, 2014, https://permaculturenews.org/2014/02/28/bamboo-drip-irrigation.
27. Art Ludwig, *Water Storage: Tanks, Cisterns, Aquifers, and Ponds for Domestic Supply, Fire, and Emergency Use*, 2nd edition (n.p.: Oasis Design, 2009).
28. Sepp Holzer, *Desert or Paradise: Restoring Endangered Landscapes Using Water Management, Including Lake and Pond Construction* (White River Junction, VT: Chelsea Green, 2012).

**Chapter Seven: Woodlands and Wastes**

1. M. Kat Anderson, *Tending the Wild: Native American Knowledge and the Management of California's Natural Resources* (Berkeley, Los Angeles, London: University of California Press, 2005), 3.
2. James C. Scott, *Seeing Like a State: How Certain Schemes to Improve the Human Condition Have Failed* (New Haven and London: Yale University Press, 1998), 12.
3. Scott, *Seeing Like a State*, 13.
4. Anderson, *Tending the Wild*, 42–43.
5. Simon Fairlie, "The Tragedy of the Tragedy of the Commons," *Dark Mountain* 1 (Summer 2010).
6. Karl Jacoby, *Crimes Against Nature: Squatter, Poachers, Thieves, and the Hidden History of American Conservation* (Berkeley, Los Angeles, London: University of California Press, 2003); Mark David Spence, *Dispossessing the Wilderness: Indian Removal and the Making of the National Parks*, revised edition (New York and London: Oxford University Press, 2000).
7. Ben Law, *The Woodland Way: A Permaculture Approach to Sustainable Woodland Management*, revised and updated (White River Junction, VT: Chelsea Green, 2013).

8. J. Russell Smith, *Tree Crops: A Permanent Agriculture* (Greenwich, CT: Devon-Adair, 1950); Wendell Berry, "Simple Solutions, Package Deals, and a 50-Year Farm Bill," in Wendell Berry, *What Matters? Economics for a Renewed Commonwealth* (Berkeley: Counterpoint, 2010), 55–69; Mark Shepard, *Restoration Agriculture: Real-World Permaculture for Farmers* (Austin, TX: Acres USA, 2013).
9. For two examples, see Wendell Berry's essay "A Forest Conversation," in *Our Only World* (Berkeley: Counterpoint, 2015), 21–52; and "Listening to the Forest," on Menominee forest management in Wisconsin, in Ted Bernard and Jora Young, *The Ecology of Hope: Communities Collaborate for Sustainability* (Gabriola Island, BC: New Society, 1997), 93–112.
10. Ianto Evans, Michael G. Smith, and Linda Smiley, *The Hand-Sculpted House: A Practical and Philosophical Guide to Building a Cob Cottage* (White River Junction, VT: Chelsea Green, 2002), 258–72.
11. Carolyn Merchant, *The Death of Nature: Women, Ecology and the Scientific Revolution* (San Francisco: Harper and Row, 1980), 170–71. For the traditional view, still alive in the work of Milton, see Wendell Berry, "The Presence of Nature in the Natural World: A Long Conversation," in *The Art of Loading Brush: New Agrarian Writings* (Berkeley: Counterpoint, 2017), 103–78.
12. Stephen Harrod Buhner, *The Lost Language of Plants: The Ecological Importance of Plant Medicines to Life on Earth* (White River Junction, VT: Chelsea Green, 2002), 80–82.
13. David Lyle, *The Book of Masonry Stoves: Rediscovering an Old Way of Warming* (White River Junction, VT: Chelsea Green, 1984), 65–68.
14. Lyle, *Book of Masonry Stoves*, 70–80, 88–97.
15. Lyle, *Book of Masonry Stoves*, 36.
16. Ianto Evans and Leslie Jackson, *Rocket Mass Heaters*, 3rd edition (Coquille, OR: Cob Cottage, 2013).
17. Kiko Denzer, with Hannah Field, *Build Your Own Earth Oven*, 3rd edition (Blodgett, OR: Hand Print Press, 2007).
18. Anderson, *Tending the Wild*, xvi.
19. See, for example, Monte Burch, *Wildlife and Woodlot Management: A Comprehensive Handbook for Food Plot and Habitat Development* (New York: Skyhorse Publishing, 2013).

**Chapter Eight: It Takes a Village: Leisure, Community, and Resilience**

1. Marshall Sahlins, *Stone Age Economics* (Chicago: Aldine, 1972), chapter 1.
2. Pierre Clastres, *Society Against the State: Essays in Political Anthropology* (New York: Zone Books, 1987), 191.

3. Clastres, *Society Against the State*, 196.
4. Juliet B. Schor, *The Overworked American: The Unexpected Decline of Leisure* (New York: Basic Books, 1993), 46–47.
5. Barbara Ehrenreich, *Dancing in the Streets: A History of Collective Joy* (New York: Henry Holt, 2006), 101.
6. Frederick W. Turner, *Rediscovering America: John Muir in His Time and Ours* (New York: Viking, 1985).
7. Karl Polanyi, *The Great Transformation: The Political and Economic Origins of Our Time* (Boston: Beacon, 2001), 81–107.
8. Douglas Hay et al., *Albion's Fatal Tree: Crime and Society in Eighteenth-Century England* (New York: Random House, 1976).
9. Clastres, *Society Against the State*, 193.
10. Ehrenreich, *Dancing in the Streets*, 23–24, referring to the work of Robin Dunbar, *Grooming, Gossip, and the Evolution of Language* (Cambridge, MA: Harvard University Press, 1996).
11. Lawrence Goodwyn, *The Populist Moment: A Short History of the Agrarian Revolt in America* (Oxford, London, New York: Oxford University Press, 1978), 20–35.
12. Linus Pauling et al., *The Triple Revolution* (Santa Barbara, CA: Ad Hoc Committee on the Triple Revolution, 1964).
13. Les Leopold, *Runaway Inequality: An Activist's Guide to Economic Justice*, 2nd edition (New York: Labor Institute Press, 2015).
14. John Michael Greer, *Dark Age America: Climate Change, Cultural Collapse, and the Hard Future Ahead* (Gabriola Island, BC: New Society, 2016), 106–9.
15. Johnny Cash, vocalist, "Sunday Morning Coming Down," lyrics by Kris Kristofferson, *The Essential Johnny Cash*, Sony, 1998. Originally recorded in 1969 and released in May 1970.
16. Ehrenreich, *Dancing in the Streets*, 24.
17. David Fleming, *Surviving the Future: Culture, Carnival and Capital in the Aftermath of the Market Economy* (White River Junction, VT: Chelsea Green, 2016), 53–71.
18. David Graeber, *Debt: The First 5,000 Years* (Brooklyn, NY, and London: Melville House, 2011). On the myth of barter, see chapter 2.
19. Sahlins, *Stone Age Economics,* 186.
20. Fleming, *Surviving the Future*, 17.
21. Graeber, *Debt*, 29–34.
22. George M. Foster, "Peasant Society and the Image of Limited Good," *American Anthropologist New Series* 67, no. 2 (1965): 293–315.
23. James C. Scott, *The Moral Economy of the Peasant: Rebellion and Subsistence in Southeast Asia* (New Haven and London: Yale

University Press, 1976); and James C. Scott, *Weapons of the Weak: Everyday Forms of Peasant Resistance* (New Haven and London: Yale University Press, 1985).

24. Frank M. Bryan, *Real Democracy: The New England Town Meeting and How It Works* (Chicago and London: University of Chicago Press, 2004); and Frank Bryan and John McClaughry, *The Vermont Papers: Recreating Democracy on a Human Scale* (White River Junction, VT: Chelsea Green, 1989).
25. Elinor Ostrom, *Governing the Commons: The Evolution of Institutions for Collective Action* (New York and Cambridge, U.K.: Cambridge University Press, 1990), 61–69.
26. Ostrom, *Governing the Commons*, 69–88.
27. New Mexico Acequia Association, "Communal Water Sovereignty: Acequias in New Mexico," *The New Farmer's Almanac 2017* (Greenhorns/Versa Press, 2017), 68–70.
28. Ostrom, *Governing the Commons*, 144–57.
29. New Mexico Acequia Association, "Communal Water Sovereignty."
30. Ostrom, *Governing the Commons*, 61–88, 173–78.
31. Didi Pershouse, *The Ecology of Care: Medicine, Agriculture, Money, and the Quiet Power of Human and Microbial Communities* (Thetford Center, VT: Mycelium Books, 2016), 242.
32. The Ithaca Health Alliance, founded in 1997, has a stable membership of 700, with 150 providers from allopathic MDs to herbalists and doulas. It helped establish a free clinic with funds collected from its membership, offering multidisciplinary holistic services.
33. Interview with John Ikerd, Local Food Summit 2017, August 7, 2017, http://www.localfoodsummit.com.
34. Peter Kalmus, "A Radical Vision for Food: Everyone Growing It for Each Other," *Truthout*, January 1, 2018, http://www.truth-out.org/news/item/43090-a-radical-vision-for-food-everyone-growing-it-for-each-other.
35. SHED, "Severine von Tscharner Fleming and Greenhorns at SHED," *Blog: Notes from the Field*, March 12, 2014, https://healdsburgshed.com/2014/03/12/severine-von-tscharner-fleming-greenhorns-shed.
36. Kirkpatrick Sale, *Human Scale Revisited: A New Look at the Classic Case for a Decentralist Future* (White River Junction, VT: Chelsea Green, 2017), 130.
37. Thomas A. Lyson, *Civic Agriculture: Reconnecting Farm, Food, and Community* (Medford, MA: Tufts University Press, 2004), 64–68.
38. Dar Williams, *What I Found in a Thousand Towns: A Traveling Musician's Guide to Rebuilding America's Communities—One Coffee Shop, Dog Run, and Open-Mike Night at a Time* (New York: Basic Books, 2017).

**Chapter Nine: Getting a Living, Forging a Livelihood**

1. Wendell Berry, *It All Turns on Affection: The Jefferson Lecture and Other Essays* (Berkeley: Counterpoint, 2012), 37.
2. Smaje, "Three Stories, Many Variables."
3. Wendell Berry, "The Making of a Marginal Farm" (1980), reprinted in *The World-Ending Fire: The Essential Wendell Berry* (Allen Lane, 2017), 44.
4. Wendell Berry, "The Agrarian Standard," reprinted in *The World-Ending Fire*, 143.
5. David Fleming, *Surviving the Future: Culture, Carnival and Capital in the Aftermath of the Market Economy* (White River Junction, VT: Chelsea Green, 2016), 15–17.
6. "The Greenhorns Documentary," 2009.
7. Wendell Berry, "Horse-Drawn Tools and the Doctrine of Labor Saving," in *The World-Ending Fire*, 158.
8. Gene Logsdon, *Letter to a Young Farmer: How to Live Richly Without Wealth on the New Garden Farm* (White River Junction, VT: Chelsea Green, 2017), 50.
9. Logsdon, *Letter to a Young Farmer*, 51.
10. Stuart Brand, *How Buildings Learn: What Happens After They're Built* (New York: Penguin Books, 1995).
11. Logsdon, *Letter to a Young Farmer*, 52, 53.
12. Ben Hartman, *The Lean Farm Guide to Growing Vegetables* (White River Junction, VT: Chelsea Green, 2017), 85.
13. Mark Shepard, *Restoration Agriculture: Real-World Permaculture for Farmers* (Austin, TX: Acres USA, 2013), 82–87.
14. Chris Smaje does the numbers and estimates that very small farm agriculture could feed up to 150 percent of England's current population. Chris Smaje, "Three Acres and a Cow," *Small Farm Future* (blog), https://smallfarmfuture.org.uk/category/neo-peasant-wessex. See also George Monbiot, "Peasant Farmers Offer the Best Chance of Feeding the World. So Why Do We Treat Them with Contempt?" *The Guardian*, June 10, 2008.
15. Fleming, *Surviving the Future*, 99–100.
16. Fleming, *Surviving the Future*, 16.
17. Fleming, *Surviving the Future*, 17.
18. Berry, *It All Turns on Affection*, 31.
19. Gene Logsdon, *Letter to a Young Farmer*, 16–17.
20. Wendell Berry, "The Total Economy" (2000), reprinted in *The World-Ending Fire*, 79.

21. Leopold Kohr, *The Overdeveloped Nations*, quoted in Kirkpatrick Sale, *Human Scale Revisited: A New Look at the Classic Case for a Decentralist Future* (White River Junction, VT: Chelsea Green, 2017), 129.
22. See Grace Lee Boggs, *The Next American Revolution: Sustainable Activism for the Twenty-First Century* (Berkeley, Los Angeles, London: University of California Press, 2011), chapter 4.
23. Perrine Hervé-Gruyer and Charles Hervé-Gruyer, *Miraculous Abundance: One Quarter Acre, Two French Farmers, and Enough Food to Feed the World* (White River Junction, VT: Chelsea Green, 2014), 180–92.

**Chapter Ten: Farmer, Citizen, Survivor: Politics and Resilience**

1. James C. Scott, *The Art of Not Being Governed: An Anarchist History of Upland Southeast Asia* (New Haven and London: Yale University Press, 2010), 24.
2. Scott, *The Art of Not Being Governed*, 29.
3. Scott, *The Art of Not Being Governed*, 183.
4. Oscar Handlin, *The Uprooted* (New York: Grosset and Dunlap, 1951).
5. Wendell Berry, "The Making of a Marginal Farm," reprinted in *The World-Ending Fire: The Essential Wendell Berry* (Allen Lane, 2017), 45.
6. James C. Scott, *Weapons of the Weak: Everyday Forms of Peasant Resistance* (New Haven and London: Yale University Press, 1985).
7. Wendell Berry, "The Total Economy," in *The World-Ending Fire*, 77.
8. Karl Marx, *The Eighteenth Brumaire of Louis Napoleon* (New York: International Publishers, 1963), 124. Marx's judgment pertained to nineteenth-century French peasants. Later he was to speculate that the Russian *mir* (communal village) might become the core of a Russian communist society. See Teodor Shanin, *Late Marx and the Russian Road: Marx and the Peripheries of Capitalism* (New York: Monthly Review Press, 1983). His doctrinaire follower, Vladimir Ilyich Lenin, could neither understand nor endorse this view.
9. Andro Linklater, *Owning the Earth: The Transforming History of Land Ownership* (New York and London: Bloomsbury, 2013), 125.
10. "John Ball (priest)," Wikipedia, https://en.wikipedia.org/wiki/John_Ball_(priest)#cite_note-bbc-6.
11. Linklater, *Owning the Earth*, 36–46.
12. Charles Tilly, *The Politics of Collective Violence* (Cambridge, U.K., and New York: Cambridge University Press, 2003), 44–53.
13. Eric R. Wolf, *Peasant Wars of the Twentieth Century* (New York, San Francisco, London: Harper and Row, 1969); Jeff Goodwin, *No Other Way Out: States and Revolutionary Movements, 1945–1991* (Cambridge, U.K., and New York: Cambridge University Press, 2001).

14. Scott, *Weapons of the Weak*, 29.
15. Thomas A. Woods, *Knights of the Plow: Oliver H. Kelley and the Origins of the Grange in Republican Ideology* (Ames: Iowa State University Press, 1991), 147–64.
16. Lawrence Goodwyn, *The Populist Moment: A Short History of the Agrarian Revolt in America* (Oxford, London, New York: Oxford University Press, 1978), especially chapter 9.
17. Goodwyn, *The Populist Moment*, 297, 307.
18. Author's unpublished research.
19. Severine Tscharner Fleming, "Introduction," *The New Farmer's Almanac 2017* (Greenhorns/Versa Press, 2017), 29–31.
20. See the review of the literature in Kirkpatrick Sale, *Human Scale Revisited: A New Look at the Classic Case for a Decentralist Future* (White River Junction, VT: Chelsea Green, 2017), chapter 18.
21. See Wendell Berry's recent defense of the principles that lay behind one such scheme, the New Deal–born Burley Tobacco Growers Co-operative Association, and the current indifference to any such solution among the nation's politicians, academics, and public officials, in "The Thought of Limits in a Prodigal Age," in *The Art of Loading Brush: New Agrarian Writings* (Berkeley: Counterpoint, 2017), 44–50.
22. Barrington Moore Jr., *Social Origins of Dictatorship and Democracy: Lord and Peasant in the Making of the Modern World* (Boston: Beacon Press, 1966), 453, 496–97, 505.
23. Stefan Hedlund, *Private Agriculture in the Soviet Union* (London and New York: Routledge, 1989), 25–28.
24. Elisabeth Croll, "Some Implications of the Rural Economic Reforms for the Chinese Peasant Household," in Ashwani Saith, editor, *The Re-emergence of the Chinese Peasantry: Aspects of Rural Decollectivisation* (London, New York, Sydney: Croom Helm, 1987), 105; and Alexander F. Day, *The Peasant in Postsocialist China: History, Politics, and Capitalism* (New York and Cambridge, U.K.: Cambridge University Press, 2013), 2.

# Bibliography

Anderson, Edgar. *Plants, Man and Life.* Berkeley, Los Angeles, London: University of California Press, 1971.

Anderson, M. Kat. *Tending the Wild: Native American Knowledge and the Management of California's Natural Resources.* Berkeley, Los Angeles, London: University of California Press, 2005.

Arsenault, Chris. "Only 60 Years of Farming Left if Soil Degradation Continues." *Scientific American*, 2014, https://www.scientificamerican.com/article/only-60-years-of-farming-left-if-soil-degradation-continues, accessed December 31, 2017.

Atlee, Tom. *The Tao of Democracy: Using Co-Intelligence to Create a World That Works for All.* Cranston, RI: Writers Collective, 2003.

Bernard, Ted, and Jora Young. "Listening to the Forest." In *The Ecology of Hope: Communities Collaborate for Sustainability*, 93–112. Gabriola Island, BC: New Society, 1997.

Berry, Wendell. "The Agrarian Standard." In *The World-Ending Fire: The Essential Wendell Berry*, selected and introduced by Paul Kingsnorth, 133–43. Allen Lane, 2017.

Berry, Wendell. "A Forest Conversation." In *Our Only World.* Berkeley: Counterpoint Press, 2015, 21–52.

Berry, Wendell. "Horse-Drawn Tools and the Doctrine of Labor Saving." In *The World-Ending Fire*, 152–59.

Berry, Wendell. *It All Turns on Affection: The Jefferson Lecture and Other Essays.* Berkeley: Counterpoint, 2012.

Berry, Wendell. "The Making of a Marginal Farm." In *The World-Ending Fire*, 37–47.

Berry, Wendell. "The Presence of Nature in the Natural World: A Long Conversation." In *The Art of Loading Brush: New Agrarian Writings*, 103–78. Berkeley: Counterpoint, 2017.

Berry, Wendell. "Simple Solutions, Package Deals, and a 50-Year Farm Bill." In *What Matters? Economics for a Renewed Commonwealth*, 55–69. Berkeley: Counterpoint, 2010.

Berry, Wendell. "The Thought of Limits in a Prodigal Age." In *The Art of Loading Brush: New Agrarian Writings*, 19–56. Berkeley: Counterpoint, 2017.

Berry, Wendell. "The Total Economy." In *The World-Ending Fire*, 66–81.

Berry, Wendell. *The Unsettling of America: Culture and Agriculture.* San Francisco: Sierra Club Books, 1977.

Bloch, Marc. *French Rural History: An Essay on its Basic Characteristics.* Berkeley and Los Angeles: University of California Press, 1966.

Boggs, Grace Lee. *The Next American Revolution: Sustainable Activism for the Twenty-First Century.* Berkeley, Los Angeles, London: University of California Press, 2011.

Bonsall, Will. *Will Bonsall's Essential Guide to Radical, Self-Reliant Gardening.* White River Junction, VT: Chelsea Green, 2015.

Boserup, Ester. *The Conditions of Agricultural Growth: The Economics of Agrarian Change Under Population Pressure.* Chicago: Aldine Publishing, 1965.

Bourne Jr., Joel K. *The End of Plenty: The Race to Feed a Crowded World.* New York: W. W. Norton, 2015.

Brand, Stuart. *How Buildings Learn: What Happens After They're Built.* New York: Penguin Books, 1995.

Bryan, Frank M. *Real Democracy: The New England Town Meeting and How It Works.* Chicago and London: University of Chicago Press, 2004.

Bryan, Frank, and John McClaughry. *The Vermont Papers: Recreating Democracy on a Human Scale.* White River Junction, VT: Chelsea Green, 1989.

Bubel, Mike, and Nancy Bubel. *Root Cellaring: Natural Cold Storage of Fruits and Vegetables*, 2nd edition. North Adams, MA: Storey Publishing, 1991.

Buhner, Stephen Harrod. *The Lost Language of Plants: The Ecological Importance of Plant Medicines to Life on Earth.* White River Junction, VT: Chelsea Green, 2002.

Burch, Monte. *Wildlife and Woodlot Management: A Comprehensive Handbook for Food Plot and Habitat Development.* New York: Skyhorse Publishing, 2013.

Christian, David. *Maps of Time: An Introduction to Big History.* Berkeley, Los Angeles, London: University of California Press, 2004.

Clastres, Pierre. *Society Against the State: Essays in Political Anthropology.* New York: Zone Books, 1987.

Coleman, Eliot. *Four-Season Harvest: Organic Vegetables from Your Home Garden All Year Long.* White River Junction VT: Chelsea Green, 1999.

Coleman, Eliot. *The New Organic Grower: A Master's Manual of Tools and Techniques for the Home and Market Gardener*, revised and expanded edition. White River Junction, VT: Chelsea Green, 1995.

Courtois-Gérard, Joseph. *Practical Handbook of Market Gardening*, 6th edition. Translated by Carol Cox. Willits, CA: self-published by translator, November 2017.

Crews, Carole. *Clay Culture: Plasters, Paints and Preservation.* Ranchos de Taos, NM: Gourmet Adobe Press, 2010.

Croll, Elisabeth. "Some Implications of the Rural Economic Reforms for the Chinese Peasant Household." In *The Re-emergence of the Chinese Peasantry: Aspects of Rural Decollectivisation*, edited by Ashwani Saith. London, New York, Sydney: Croom Helm, 1987.

Day, Alexander F. *The Peasant in Postsocialist China: History, Politics, and Capitalism.* New York and Cambridge, U.K.: Cambridge University Press, 2013.

De Decker, Kris. "Fruit Walls: Urban Farming in the 1600s." *Low-Tech Magazine*, http://www.lowtechmagazine.com/2015/12/fruit-walls-urban-farming.html.

Denzer, Kiko, with Hannah Field. *Build Your Own Earth Oven*, 3rd edition. Blodgett, OR: Hand Print Press, 2007.

Deppe, Carol. *Breed Your Own Vegetable Varieties: The Gardener's and Farmer's Guide to Plant Breeding and Seed Saving.* White River Junction, VT: Chelsea Green, 2000.

Deppe, Carol. *The Resilient Gardener.* White River Junction, VT: Chelsea Green, 2010.

Deppe, Carol. *The Tao of Vegetable Gardening: Cultivating Tomatoes, Greens, Peas, Beans, Squash, Joy, and Serenity.* White River Junction, VT: Chelsea Green, 2015.

Donahue, Brian. *Reclaiming the Commons: Community Farms and Forests in a New England Town.* New Haven and London: Yale University Press, 1999.

Ehrenreich, Barbara. *Dancing in the Streets: A History of Collective Joy.* New York: Henry Holt, 2006.

Evans, Ianto, and Leslie Jackson. *Rocket Mass Heaters*, 3rd edition. Coquille, OR: Cob Cottage Company, 2013.

Evans, Ianto, Michael G. Smith, and Linda Smiley. *The Hand-Sculpted House: A Practical and Philosophical Guide to Building a Cob Cottage.* White River Junction, VT: Chelsea Green, 2002.

Fairlie, Simon. *Low Impact Development: Planning and People in a Sustainable Countryside.* Charlbury, Oxfordshire, U.K.: Jon Carpenter, 1996.

Fairlie, Simon. "The Tragedy of the Tragedy of the Commons." *Dark Mountain* 1 (Summer 2010), 178–200.

Falk, Ben. *The Resilient Farm and Homestead.* White River Junction, VT: Chelsea Green, 2013.

Fitzgerald, Deborah. *Every Farm a Factory: The Industrial Ideal in American Agriculture.* New Haven and London: Yale University Press, 2003.

Fleming, David. *Surviving the Future: Culture, Carnival and Capital in the Aftermath of the Market Economy.* White River Junction, VT: Chelsea Green, 2016.

Fortier, Jean-Martin. *The Market Gardener.* Gabriola Island, BC: New Society, 2014.

Foster, George M. "Peasant Society and the Image of the Limited Good." *American Anthropologist New Series* 67, no. 2 (April 1965), 293–315.

Gates, Paul Wallace. "The Homestead Law in an Incongruous Land System." In *The People's Land: A Reader on Land Reform in the United States*, edited by Peter Barnes, 7–11. Emmaus, PA: Rodale Press, 1975.

*Global Water Security: Intelligence Community Assessment*, 2012, https://www.dni.gov/files/documents/Special%20Report_ICA%20Global%20Water%20Security.pdf.

Goodwin, Jeff. *No Other Way Out: States and Revolutionary Movements, 1945–1991.* Cambridge, U.K., and New York: Cambridge University Press, 2001.

Goodwyn, Lawrence. *The Populist Moment: A Short History of the Agrarian Revolt in America.* Oxford, London, New York: Oxford University Press, 1978.

Graber, Cynthia. "Farming Like the Incas." Smithsonian.com, September 6, 2011, https://www.smithsonianmag.com/history/farming-like-the-incas-70263217.

Graeber, David. *Debt: The First 500 Years.* Brooklyn, NY: Melville House, 2011.

Greer, John Michael. *Dark Age America: Climate Change, Cultural Collapse, and the Hard Future Ahead.* Gabriola Island, BC: New Society Publishes, 2016.

Greer, John Michael. *The EcoTechnic Future: Envisioning a Post-Peak World.* Gabriola Island, BC: New Society, 2009.

Greer, John Michael. *The Long Descent: A User's Guide to the End of the Industrial Age.* Gabriola Island, BC: New Society, 2008.

Guthman, Julie. *Agrarian Dreams: The Paradox of Organic Farming in California.* Berkeley, Los Angeles, London: University of California Press, 2004.

Handlin, Oscar. *The Uprooted.* New York: Grosset and Dunlap, 1951.

Hartman, Ben. *The Lean Farm.* White River Junction, VT: Chelsea Green, 2015.

Hartman, Ben. *The Lean Farm Guide to Growing Vegetables.* White River Junction, VT: Chelsea Green, 2017.

Hay, Douglas, et al. *Albion's Fatal Tree: Crime and Society in Eighteenth-Century England.* New York: Random House, 1976.

Healy, Kevin. *Llamas, Weavings, and Organic Chocolate: Multicultural Grassroots Development in the Andes and Amazon of Bolivia.* South Bend, IN: University of Notre Dame Press, 2001.

Hedlund, Stefan. *Private Agriculture in the Soviet Union.* London and New York: Routledge, 1989.

Heinberg, Richard. *The Party's Over: Oil, War and the Fate of Industrial Societies*, revised and updated edition. Gabriola Island, BC: New Society, 2005.

Heinberg, Richard, and David Findley, *Our Renewable Future: Laying the Path for One Hundred Percent Clean Energy.* Washington DC: Island Press, 2016.

Hervé-Gruyer, Perrine, and Charles Hervé-Gruyer. *Miraculous Abundance: One Quarter Acre, Two French Farmers, and Enough Food to Feed the World.* White River Junction, VT: Chelsea Green, 2016.

Holzer, Sepp. *Desert or Paradise: Restoring Endangered Landscapes Using Water Management, Including Lake and Pond Construction.* White River Junction, VT: Chelsea Green, 2012.

Hyams, E. *Soil and Civilization.* London and New York: Thames and Hudson, 1952.

Jacoby, Karl. *Crimes Against Nature: Squatter, Poachers, Thieves, and the Hidden History of American Conservation.* Berkeley, Los Angeles, London: University of California Press, 2003.

Johnson, Warren. *Muddling Toward Frugality.* San Francisco: Sierra Club Books, 1978.

Kalmus, Peter. "A Radical Vision for Food: Everyone Growing It for Everyone." *Truthout*, January 1, 2018. http://www.truth-out.org/news/item/43090-a-radical-vision-for-food-everyone-growing-it-for-each-other.

Katazawa, Satoshi. "Common Misconceptions About Science II: Life Expectancy." *Psychology Today*, November 20, 2008. https://www.psychologytoday.com/us/blog/the-scientific-fundamentalist/200811/common-misconceptions-about-science-ii-life-expectancy.

Katz, Sandor Ellix. *The Art of Fermentation.* White River Junction, VT: Chelsea Green, 2012.

Kimbrell, Andrew, editor. *Fatal Harvest: The Tragedy of Industrial Agriculture.* Washington, DC, Covelo, CA, and London: Island Press, 2002.

King, F. H. *Farmers of Forty Centuries: Organic Farming in China, Korea, and Japan.* Mineola, NY: Dover Publications, 2004 [1911].

Krishnamurthy, Ravindra. "Bamboo Drip Irrigation." Permaculture News. https://permaculturenews.org/2014/02/28/bamboo-drip-irrigation.

Law, Ben. *The Woodland Way: A Permaculture Approach to Sustainable Woodland Management*, revised and updated. White River Junction, VT: Chelsea Green, 2013.

Lengnick, Laura. *Resilient Agriculture: Cultivating Food Systems for a Changing Climate.* Gabriola Island, BC: New Society, 2015.

Leopold, Les. *Runaway Inequality: An Activist's Guide to Economic Justice*, 2nd edition. The Labor Institute, 2015.

Lightfoot, Dale R. "Syrian Qanat Romani." WaterHistory.org. http://waterhistory.org/histories/syria.

Linklater, Andro. *Owning the Earth: The Transforming History of Land Ownership.* New York, London, New Delhi, Sydney: Bloomsbury, 2013.

Logsdon, Gene. *Letter to a Young Farmer: How to Live Richly Without Wealth on the New Garden Farm.* White River Junction, VT: Chelsea Green, 2017.

Lowdermilk, W. C. *Conquest of the Land Through 7,000 Years.* US Department of Agriculture, Soil Conservation Service, Agriculture Information Bulletin no. 99. Washington, DC: Government Printing Office, 1953.

Ludwig, Art. *Water Storage: Tanks, Cisterns, Aquifers, and Ponds for Domestic Supply, Fire, and Emergency Use*, 2nd edition. N.p.: Oasis Design, 2009.

Lyle, David. *The Book of Masonry Stoves: Rediscovering an Old Way of Warming.* White River Junction, VT: Chelsea Green, 1984.

Lyson, Thomas A. *Civic Agriculture: Reconnecting Farm, Food, and Community.* Medford, MA: Tufts University Press, 2004.

MacDonald, Angus. "The Family Farm Is the Most Efficient Unit of Production." In *The People's Land: A Reader on Land Reform in the United States*, edited by Peter Barnes, 86–88. Emmaus, PA: Rodale Press, 1975.

Madison, Mike. *Fruitful Labor: The Ecology, Economy, and Practice of a Family Farm.* White River Junction, VT: Chelsea Green, 2018.

Mahoney, James. *The Legacies of Liberalism: Path Dependence and Political Regimes in Central America.* Baltimore: Johns Hopkins University Press, 2001.

Mann, Charles C. *1491: New Revelations of the Americas Before Columbus.* New York: Knopf, 2005.

Mann, Charles C. *1493: Uncovering the New World Columbus Created.* New York: Alfred A. Knopf, 2011.

Marsh, George Perkins. *Man and Nature; or, Physical Geography as Modified by Human Action.* New York: Charles Scribner, 1864.

Marx, Karl. *The Eighteenth Brumaire of Louis Bonaparte.* New York: International Publishers, 1963.

Mayhew, Alan. *Rural Settlement and Farming in Germany.* London: B. T. Batsford, 1973.

Merchant, Carolyn. *The Death of Nature: Women, Ecology and the Scientific Revolution.* San Francisco: Harper and Row, 1980.

Monbiot, George. "Peasant Farmers Offer the Best Chance of Feeding the World. So Why Do We Treat Them with Contempt?" *The Guardian*, June 10, 2008.

Monbiot, George. "We're Treating the Soil Like Dirt," *The Guardian*, March 25, 2015, https://www.theguardian.com/commentisfree/2015/mar/25/treating-soil-like-dirt-fatal-mistake-human-life.

Montgomery, David R. *Dirt: The Erosion of Civilizations.* Berkeley, Los Angeles, London: University of California Press, 2007.

Montgomery, David R. *Growing a Revolution: Bringing Our Soil Back to Life.* New York and London: W. W. Norton, 2017.

Montgomery, David R., and Anne Biklé. *The Hidden Half of Nature: The Microbial Roots of Life and Health.* New York and London: W. W. Norton, 2016.

Nabhan, Gary Paul. *Enduring Seeds: Native American Agriculture and Wild Plant Conservation.* Tucson: University of Arizona Press, 1989.

Nabhan, Gary Paul. *Growing Food in a Hotter, Drier Land: Lessons from Desert Farmers on Adapting to Climate Uncertainty.* White River Junction, VT: Chelsea Green, 2013.

New Mexico Acequia Association. "Communal Water Sovereignty: Acequias in New Mexico." In *The New Farmer's Almanac 2017*, 68–70. Greenhorns/Versa Press, 2017.

Newman, Chris. "Yes, Organic Farming Will Kill Us All," https://shift.newco.co/yes-organic-farming-will-kill-us-all-12d900979cf2.

Oppenheimer, Todd. "The Drought Fighter." *Craftsmanship Quarterly*, January 2015, https://craftsmanship.net/drought-fighters.

Ostrom, Elinor. *Governing the Commons: The Evolution of Institutions for Collective Action.* New York and Cambridge, U.K.: Cambridge University Press, 1990.

Pauling, Linus, et al. *The Triple Revolution.* Santa Barbara, CA: Ad Hoc Committee on the Triple Revolution, 1964.

Pearse, C. Kenneth. "Qanats in the Old World: Horizontal Wells in the New." *Journal of Range Management* 265 (September 1973), 320–21.

Pershouse, Didi. *The Ecology of Care: Medicine, Agriculture, Money, and the Quiet Power of Human and Microbial Communities.* Thetford Center, VT: Mycelium Books, 2016.

Phillips, Michael. *The Apple Grower: A Guide for the Organic Orchardist*, 2nd edition. White River Junction, VT: Chelsea Green, 2005.

Phillips, Michael. *The Holistic Orchard: Tree Fruits and Berries the Biological Way.* White River Junction, VT: Chelsea Green, 2012.

Polanyi, Karl. *The Great Transformation: The Political and Economic Origins of Our Time.* Boston: Beacon, 2001.

Rappaport, Roy A. *Pigs for the Ancestors: Ritual in the Ecology of a New Guinea People*, revised edition. New Haven: Yale University Press, 1984.

Richards, Paul. *Indigenous Agricultural Revolution.* Boulder, CO: Westview Press, 1985.

Sahlins, Marshall. *Stone Age Economics.* Chicago: Aldine, 1972.

Sahrawat, K. L. "Fertility and Organic Matter in Submerged Rice Soils." *Current Science* 88, no. 5 (March 10, 2005).

Sale, Kirkpatrick. *Human Scale Revisited: A New Look at the Classic Case for a Decentralist Future.* White River Junction, VT: Chelsea Green, 2017.

Schor, Juliet B. *The Overworked American: The Unexpected Decline of Leisure.* New York: Basic Books, 1993.

Scott, James C. *Against the Grain: A Deep History of the Earliest States.* New Haven and London: Yale University Press, 2017.

Scott, James C. *The Art of Not Being Governed: An Anarchist History of Upland Southeast Asia.* New Haven and London: Yale University Press, 2010.

Scott, James C. *The Moral Economy of the Peasant: Rebellion and Subsistence in Southeast Asia.* New Haven and London: Yale University Press, 1976.

Scott, James C. *Seeing Like a State: How Certain Schemes to Improve the Human Condition Have Failed.* New Haven and London: Yale University Press, 1998.

Scott, James C. *Weapons of the Weak: Everyday Forms of Peasant Resistance.* New Haven and London: Yale University Press, 1985.

Shanin, Teodor. *Late Marx and the Russian Road: Marx and the Peripheries of Capitalism.* New York: Monthly Review Press, 1983.

SHED. "Severine von Tscharner Fleming and Greenhorns at SHED." *Blog: Notes from the Field.* March 12, 2014. https://healdsburgshed.com/2014/03/12/severine-von-tscharner-fleming-greenhorns-shed.

Shepard, Mark. *Restoration Agriculture: Real-World Permaculture for Farmers.* Austin, TX: Acres USA, 2013.

Smaje, Chris. "The Return of the Peasant. Or, the History of the World in Ten-and-a-Half Blog Posts." *Small Farm Future* (blog), https://smallfarmfuture.org.uk/research/the-return-of-the-peasant-or-the-history-of-the-world-in-ten-and-a-half-blog-posts.

Smaje, Chris. "Three Acres and a Cow." *Small Farm Future* (blog), https://smallfarmfuture.org.uk/category/neo-peasant-wessex.

Smaje, Chris. "Three Stories, Many Variables: The Case of Small Farm Productivity." *Statistics Views*, March 5, 2015, http://www.statisticsviews

.com/details/feature/7553541/Three-stories-many-variables-the-case-of-small-farm-productivity.html.

Smith, J. Russell. *Tree Crops: A Permanent Agriculture.* Greenwich, CT: Devon-Adair, 1950.

Spence, Mark David. *Dispossessing the Wilderness: Indian Removal and the Making of the National Parks*, revised edition. New York, London: Oxford University Press, 2000.

Thomas Jr., William L., editor. *Man's Role in Shaping the Face of the Earth.* Chicago: University of Chicago Press, 1956.

Tilly, Charles. *The Politics of Collective Violence.* Cambridge, U.K., and New York: Cambridge University Press, 2003.

Tscharner Fleming, Severine von. "Introduction." In *The New Farmer's Almanac 2017*, 17–38.

Turner, Frederick W. *Rediscovering America: John Muir in His Time and Ours.* New York: Viking, 1985.

Volk, Anthony A., and Jeremy Atkinson. "Is Child Death the Crucible of Human Evolution?" *Journal of Cultural and Evolutionary Psychology*, December 2008. Proceedings of the 2nd Annual Meeting of the North-eastern Evolutionary Psychology Society.

White, Courtney. *Grass, Soil, and Hope: A Journey Through Carbon Country.* White River Junction, VT: Chelsea Green, 2013.

Wilken, Gene C. *Good Farmers: Traditional Agricultural Resource Management in Mexico and Central America.* Berkeley, Los Angeles, London: University of California Press, 1987.

Williams, Dar. *What I Found in a Thousand Towns: A Traveling Musician's Guide to Rebuilding America's Communities—One Coffee Shop, Dog Run, and Open Mic Night at a Time.* New York: Basic Books, 2017.

Wiswall, Richard. *The Organic Farmer's Business Handbook: A Complete Guide to Managing Finances, Crops, and Staff—and Making a Profit.* White River Junction, VT: Chelsea Green, 2009.

Wittfogel, Karl A. *Oriental Despotism: A Comparative Study of Total Power.* New Haven: Yale University Press, 1957.

Wolf, Eric R. *Peasant Wars of the Twentieth Century.* New York, San Francisco, London: Harper and Row, 1969.

Woods, Thomas A. *Knights of the Plow: Oliver H. Kelley and the Origins of the Grange in Republican Ideology.* Ames: Iowa State University Press, 1991.

# Index

# INDEX

# INDEX

# INDEX

INDEX

# INDEX

# INDEX

# About the Author

After twenty years in academia, Michael Foley began farming first in southern Maryland, then in Willits, California, where he, his wife, and oldest daughter currently operate Green Uprising Farm—a small, diversified farm. Foley is cofounder of the School of Adaptive Agriculture (formerly the Grange Farm School), a farmer training and education program where he is a board member and teacher. He also helped create and manage a community kitchen and small farmers group. He manages the local farmers market, and has served as vice president of the Mendocino County Farmers' Market Association and president of Little Lake Grange.